SCIENCE IS SERIOUS:
All the scientists say so
by

Dr. Gary McCallister

DEDICATION
To Dr. Ferron Andersen, who believes in late bloomers.

ACKNOWLEDGEMENTS

I would acknowledge all those people that helped with my education, except most of them don't want to be associated. My wife is my first and best editor and this book would be even more confusing and less understandable than it is. She is the miracle of my life.

TABLE OF CONTENTS

Chapter	Title	Page
SERIOUS INTRODUCTION TO THIS BOOK		
1.	BEGINNINGS	
2.	INTRODUCTION TO SERIOUS SCIENCE	
3.	THE SERIOUSNESS OF SERIOUS SCIENCE	
SERIOUS PHYSICAL SCIENCES		
4.	KINETIC ENERGY	
5.	ENTROPY	
6.	WORK	
7.	PRESSURE	
8.	GEOMETRY	
9.	TOPOLOGY	
10.	HANGING OUT WITH GRAVITY	
11.	RUN, DON'T WALK	
SERIOUS BIOLOGICAL SCIENCES		
12.	ORGANIC	
13.	HOMEOSTASIS	
14.	STRUCTURE AND FUNCTION	
15.	BOTANY	
16.	GREEN BANANAS	
17.	SKUNKS	
SERIOUS HEALTH SCIENCES		
18.	EXERCISING MEMORY	
19.	ARTHRITIS	
20.	WHISTLING DIXIE	
21.	HOW TO NOT DIE	
22.	HOW TECHNOLOGY MADE MY LIFE BETTER	
23.	METABOLISM	
SERIOUS IDEAS		
24.	THEORY	
25.	WRITING	
26.	NEUTRALITY	
27.	ANXIETY	
28.	AGNOTOLOGY	
29.	ODD OR EVEN	
30	GIVE YOURSELF A HAND	
SERIOUS SPIRITUALITY		

31.		THINGS AND MIND
32.		SOMETHING FISHY
33.		VALENTINES
34.		EASTER
35.		A UNIVERSAL HOLIDAY
36.		WHAT ARE PEOPLE FOR

SERIOUS CONCLUSIONS

37.		STRATEGY
38.		GRATITUDE
39.		ARROGANCE
40.		HOW THINGS WORK
41.		ENDINGS

SERIOUS POSTSCRIPT

INTRODUCTION TO THIS BOOK

Science rests upon several basic assumptions. These assumptions were embraced slowly and without real conscious thought over the period of several centuries. This is surprising to many people. It is commonly thought that science is based on conscious thought, so to find out the roots of science are somewhat unconscious is disturbing.

Those who have read my other science books understand science can be disturbing. My wife thinks science is just fine and it's me that is disturbing. That is yet another assumption that is growing without real conscious thought.

The problem is that it is very hard to pinpoint the actual invention of science. It just grew topsy-turvy over a period of time. Euclid was inventing geometry 300 years before Christ and Ptolemy created a theory of a geocentric model of the solar system 150 years before Christ.
Other advances in different fields were made in scattered places around the globe by many different people. But don't blame me because I wasn't one of them.

Most historians tend to list the discovery that earth had a magnetic field in 1600 by William Gilbert as a kind of an official beginning date of science. I have no idea why. We still don't know why lodestone attracts metal.

Anyway, science is a method of asking questions of the universe and the material world. Of course, the early scientists didn't know this. They were just curious about something. But eventually humans had to come up with some kind of definition. That is one kind of definition. There are others.

I'm not exactly sure what science is either. But it's cool to do it, and no one can really argue, whatever you're doing. Sometimes I call it tinkering.

So anyway, what are these basic assumptions that science, an unknown process with no known beginning, rest upon? It's easy.

1) There are rules that reliably govern the universe, although we don't know where these rules came from.
2) These rules are the same everywhere in the universe and for all time, since the beginning of time and will never change, although how we know this is not clear.
3) That science is essential to man's survival and is of the upmost seriousness in nature. We know this because scientists say so.

    There are always problems with assumptions, but when they lead to some success they may be justified. Science has had a pretty good track record of success, if sometimes the application has been a little rough. I mean, we have been successful at splitting the atom, but we mostly use the knowledge to threaten people.
    It's also always good to question assumptions. The assumption that there are uniform rules govern the universe seems reasonable, even if we don't always agree on the rules. Of course, people thought there were uniform rules that governed the universe before there was science also. They just assumed different rules.
    I don't mind the assumption that the rules have always been the same, although how we know that I'm not sure. We assume they are the same everywhere in the universe also. Oh yeah? Then I'd like to know what science fiction is all about. I also think it is a little cheeky to believe that the rules will never change. Prophecy is one area in which science does not have a stellar record.
    However, perhaps sciences biggest problem is that science is essential to survival and is, therefore, a very serious undertaking. In the first-place, humanity apparently survived for some time before science existed. But secondly, I know of absolutely no evidence that science must be serious. In fact, if one thumbs through almost any science journal, one will find at least some uproarious comedy.
    Generally, scientists don't want you to know that. This is because they need to convince people of the serious nature of their endeavor to get money to support their activities. Money is always a serious endeavor. Especially when it

appears that it is requiring increasing amounts for lessening returns.

Now, I don't want any of my science friends, that I used to have, to go hungry or anything. But I feel it is my duty to dispel the last basic assumption of science. So, in this book I present scientific topics as if they were not serious. This is not as easy as it sounds and a great deal of effort has gone into it. However, I want to assure you that, despite the difficulty, there is something true and scientific in every chapter. I think.

1 - BEGINNING

I have no idea why something begins. If you think about it, which I try not to do because I don't know how to get started, there is always something that happened before whatever is beginning that started the thing that is being begun. Like, what prompted me to begin writing about beginnings?

Here is another example. Do I awake each morning because the alarm goes off or because I set the alarm the night before? Why do I set the alarm the night before? My wife often asks why I set the stupid alarm at all. She hates alarms. But I set the alarm the night before so I will be on time to work. My department chair decided what time my first class should, start so it is the department chairs fault. He decided my first class should begin at 7:00 AM because he hates me.

So, what begins a beginning? Beginnings are especially important in science. Why does a person make the seemingly illogical decision to become a scientist? Do they think they are going to get rich or earn a Nobel Prize?

To become a scientist, you know, with the union card that says Ph.D., a person must subject themselves to years and years of mastering arcane, complex, often-non-intuitive, almost-always-mathematical, obscure, concepts about a smaller and smaller piece of our world.

There are psychological theories about why someone would do this although I don't know how those theories got started either. I wonder if there are theories about why a psychologist decides to become a psychologist. That could be interesting.

This conundrum has practical implications for even those who aren't scientists because of all the fascinating things in the world there are to see, do, and be interested in, the question becomes "Why do people begin to do the things they begin to do?" I mean, some people like to sit around in the sun and fish. Other people just like to s t around in the sun. Today, people have even become interested in cooking. My mother tried to get out of cooking whenever she could.

So, how did the guy who didn't take a math class until he was in his third year of college decide to become a scientist? What was I thinking? My Dad used to ask me that a lot. . . I wonder what made him start doing that.

Why did I study parasites? Armadillos are interesting too, but I didn't feel compelled to study them. I did get a little interested in them later when I found out they are the only animals besides humans that get leprosy. But I doubt I'd have found that very interesting had I not been interested in diseases first.

I don't know why some ideas grab my attention, and others don't. I remember my major professor handing me several articles to read over the weekend to be discussed with him later. He asked me which ones I found interesting and why. One article had an almost "throwaway line" about how the nematode parasite in one trial stopped producing viable eggs after a certain treatment. It wasn't even the point of the article, but I thought it was an interesting occurrence. Why? That very experiment took me into parasites.

Well, I couldn't very well tell him that none of them were really all that interesting. I wanted into his research program and he must have thought there was something interesting in at least one of the papers.

Of course, I had just spent two years in the US Army, which in all its wisdom, had made an English Literature major into a medical corpsman. Being a medical corpsman got me interested in science. I was in the Army after being drafted during the Viet Nam war era. So this "Simply Science" column is all the fault of President Lyndon Johnson. He meant well, I suppose.

Just to confuse things a little more, the end of something is always the beginning of something else. We've heard of the "beginning of the end." Have you ever heard of the "end of the beginning"?

Like, why did you start reading this book? Did you REALLY think you were going to learn some science? My wife asks me the same question all the time. "Who reads this stuff?" But then, she must hear a lot of it. I guess this is an unfortunate ending for a chapter on science. How did I get started on this topic anyway?

2 - INTRODUCTION TO SERIOUS SCIENCE

I have been trying to get back in touch with reality. Unfortunately, he didn't leave a forwarding address. For all I know, reality might be a she. Honestly, I've never actually made the individual's acquaintance. I felt bad about that until I Googled him, or her, and found out that no else seems to be in contact with reality either.

It's important for a scientist to be in touch with reality because the world is counting on scientists to tell the world what is real. It's obvious that science hasn't done such a good job. For example, it is not real to suppose that someone else will always be there to supply you with money, food, water, transportation, warmth, and shelter. It is nice that so many of these things are available for such a large population as ours, but having these things is not obligatory. As Stephen Crane once wrote:

A man said to the universe:
"Sir, I exist!"
"However," replied the universe,
"The fact has not created in me
A sense of obligation."

Reality is something like "The state of things as they actually exist, rather than as they may appear or are imagined." Personally, I think reality is highly overrated. I find love is far more important.

There are dozens of books on reality. Most of them have to do with magic, philosophy, computers, economics, government, and pornography: all of which are distinctly detached from reality. Like Richard Dawkins': "The Magic of Reality: How We Know What's Really True." Personally, I think that title is an unfortunate choice. Perhaps not, considering that it's Dawkins.

Another book claimed, "The goal of molecular structures is to plant the seeds of insight rather than illusion." Another, "Meditation makes the entire nervous system go into a field of coherence." I understand these statements even less than I understand reality. I rejected the book where the reviewer said the book "was 'punished' more than four months ago." Obviously, it is out of date.

Science now dominates our worldview and thoughts, as if anything that isn't material isn't worth, well, anything. However, that is blatantly untrue. There are all kinds of non-material things that are truly significant. In fact, many non-material ideas are more significant than science. Wars are seldom fought over gravity or quantum mechanics but are routinely fought over things like freedom, love, or political ideologies. I guess gravity and quantum mechanics can contribute to war once something non-material has started one.

Now, I have made my living as a scientist, and I find the occupation to be pleasant and interesting. However, I am not sure that explaining love as a hormonal attraction, or that defining religion as a genetic defect of neural connections in the brain, really adds all that much to our understanding of reality.

For that matter, the difference between a republic and democracy seems far more significant to my life than the tiny machines being built on a molecular level that received the Nobel Prize in Chemistry recently. And I maintain that political science isn't a science at all, although I still have friends who are political scientists. I'd tell you who they are, but they don't want to be associated.

There seems to be a question in some people's minds about what reality is. They worry that what we think is real is only a dream or something. My experience has been that those questions, while interesting, largely disappear when a band saw rips into your left thumb.

So, what are dreams if they occur in real people, in real time, but aren't real? Science would say that dreams can't become reality, and someone who thinks they do is not in touch with reality. This explains my quest to get back in touch with reality because it turns out that my dream became a reality fifty years ago, on October 14, 1966, when my wife married me in Rossfeld, Germany.

3 - THE SERIOUSNESS OF SERIOUS SCIENCE

What do you get when you cross a joke with a rhetorical question? I don't know either, but, what if reality is real? That's a rhetorical question, but the answer is no joke. The consequences are staggering! I just never thought of it before.

I guess, unconsciously, I have always assumed that my world was only partly real. I have spent a lot of time studying the real part, but there has always been an unreal part that was very real to me. My wife thinks this has been a problem. She just doesn't understand me.

When I woke up this morning, I found the world was pretty much like I left it. While a little discouraging, it was also comforting in a way. I was in the same room I went to sleep in. The news hadn't changed much at all for a couple of weeks. I was pretty sure I was still me, although who that was has been questioned at times. Today I still think I know what has happened previously, and I still don't have a clue what will happen tomorrow. It all seems real.

Maybe understanding reality depends on what the definition of "real" is. The verb "is" is by definition "real", isn't it? Once, in blind haste, I kicked a chair hard. After that I decided that real is whatever I can sense with my five senses.

Then someone pointed out electrons and the number "5". I cannot sense them, but they seem real enough. Oh, and then, there's déjà vu, which seems real enough, but maybe isn't. Hmmm. Have I written that before?

What I am actually kicking when I kick a chair? Chairs seem like a solid take on reality, but they quickly start to feel insubstantial from a scientific perspective. If you take a chair apart, you'll find that its basic constituent are atoms. In turn, Atoms are composed of smaller subatomic particles, themselves built of yet smaller particles.

Science needs remarkably few ingredients to account for chairs. All that is needed is a handful of different atoms, the forces that govern their interactions, plus some rules laid down by quantum mechanics. They tell me the chair is mostly empty space, the particles held apart by magnetic forces.   Tell that to my toe.

The main problem with reality, as I see it, is that if reality is real then I am pretty much stuck in it.   I may not like it very much, but there isn't much I can do about it.   It's not like its validity is determined by a Supreme Court decision or something.

Like maybe your old man beats you every night. That's not a reason to drop out of school.   In fact, that's a pretty good reason to stay in school, and find a way to do it well.   Best ticket out.   Or maybe your wife doesn't understand you.   That doesn't make it a good idea to spend more time away from home.   In fact, maybe more time at home is exactly what's needed.   In my case, my wife understands me all too well.   She kind wants me out of the house as much as possible.

When Albert Einstein finally completed his general theory of relativity in 1916, he looked at the equations and discovered an unexpected message: the universe is expanding.   He didn't like that conclusion very much and argued against it for years.   His arguments didn't matter much.   The Universe is still expanding.

Reality is so inconvenient.   It's just a fact that gravity can hurt you.   If speed doesn't kill, it at least causes a lot of typos.   Men and women must have babies if anyone is going to pay for their social security.   We would never ask what we are going to do about tornados.   Why do we think we can always control mosquitoes?   Those are rhetorical questions, but it's no joke.

SERIOUS PHYSICAL SCIENCE

Science is the study of the material world. According to the physical scientists, the physical sciences are the most important disciplines and the bedrock of all science. Of course, the mathematicians disagree. They think mathematics are the bedrock of the physical sciences. This is also odd since mathematics has little to do with material things at all. Of course, we use mathematics to manipulate material things, but math itself is mostly in the mind.

The physical sciences, namely physics and chemistry, have sought to discover the nature of the material world. They have been pretty successful, and have blessed us with electricity, mechanical systems, drugs, rockets, bombs, and other tools of mankind. I admit it is a little ambiguous since almost all of the physical sciences have caused as much pain and suffering as anything they have alleviated. They meant well, I suppose.

Anyway, it is assumed in science that one should first understand physics, then chemistry, and only then should we undertake biology or geology. I listed them in that order as alphabetical. Well, and because biology is obviously more important. Actually, most living things think biology is more important than physics or chemistry, but this is a science book so we will humor them.

The next few chapters, four through eleven to be precise, will deal with a selection of topics concerning the material world, mostly from the perspective of the physical sciences. Of course, as the reader you are in charge and if you want to skip them to get to the more important stuff, it won't make a lot of difference. The truth is you can understand a lot about the material world and life and know next to nothing about physics or chemistry. Okay, probably that is true of biology and geology as well.

4 - KINETIC ENERGY

If you put a heavy box on the top shelf of the closet, you have created potential energy. It's called potential energy because the box isn't really doing anything. But the box has the potential to fall from the shelf severely injuring your foot. There is no need for you to do this experiment. Just take my word for it.

When the box falls, it is no longer considered potential anything because it is now actually doing something, namely falling. The movement of the box is called kinetic energy. Kinetic is defined as, "of, or relating to, the movement of physical bodies." I will now use the word in a sentence. My grandchildren are very kinetic.

Physicists tell me that the amount of energy released when a box falls is exactly equal to the amount of energy that was required to lift the stupid box onto the shelf in the first place. This is one reason why I don't trust physicists very much. The chaos, damage, and pain created by the falling box obviously exceeded the energy required to lift it. But if you can suspend disbelief for a few minutes, I think I can show you why this idea is extremely important.

Let's suppose that you lift one of your kitchen table chairs onto the table. It now could potentially fall down. Next lift a second chair and place it on top of the first chair. Which chair has more potential to fall? The second one of course, because you had to lift it higher than the first chair, which took more energy. Besides, the higher the stack the more unstable it is. Now lift a third chair on top of the other two. Now what has more energy?

This is a bit of a trick question since you will guess it is the third chair for the reasons just explained. In fact, your wife will probably be the most energized as she realizes the threat to her orderly household! Disregarding your wife, it would be the third chair. Disregarding your wife, however, would not be wise. Perhaps you should do all this when she is not home.

The point is that the more things that are stacked on each other, the more energy the top part of the resultant stack represents. What if we didn't use kitchen chairs but instead stacked shoe boxes? The amount of energy in the stacked shoe boxes would be less, sure. But the third box would still possess the most energy of all the boxes.

What if we stacked nickels? The top nickel would represent the most energy, and the more nickels in the stack the more energy. Even if you stack playing cards the act of stacking would still represent the storing of energy. This may not seem very significant to you because even if a playing card fell on your foot it wouldn't hurt much.

Furthermore, the concept doesn't change when we pick up a tiny particle like a hydrogen atom and stack it onto a carbon atom. It is really no different from stacking a kitchen chair on the table, except that your wife probably won't get all mad about it. You will have still created potential energy. If you stack more than one atom onto the carbon atom, you will have created more energy than if you stack only one. A stack of hydrogen atoms on a stack of carbon atoms is otherwise known as sugar.

Now, if you knock one of the atoms off the sugar, you will release all the energy that was stored up when it was stacked. When you digest sugar, you are basically knocking atoms off carbon. And knocking hydrogen off sugar is how you got the initial energy to stack the box in the closet in the misbegotten effort at organization, which subsequently severely damaged your foot. The moral is, never waste energy getting organized.

5 - ENTROPY

How can I have a three-car garage and still must park my old truck outside? I think it must have to do with entropy. In thermodynamics, entropy is a measure of the number of specific ways in which a thermodynamic system may be arranged, and it is commonly understood as a measure of disorder.

I went to school a long time to try to understand what that meant. I still don't know, which surely says something about either our educational system or my cognitive abilities. I report. You decide.

Anyway, I think it means that because all molecules are constantly moving around due to energy in the universe, any given cluster of molecules tends to move from where they are to where they ain't. I know the correct English is "aren't", but people remember things better sometimes when stated in incorrect English. Do you suppose that's how cursing got started?

With this knowledge of thermodynamics, we can begin to understand why an empty, three-car garage eventually fills up with something besides cars. Stuff just moves to where there isn't any stuff. It's the law! Well, it's the law of thermodynamics anyway. The hotter it gets; the faster things move around and disburse into the garage.

The flip side of the law is that when things get cold, they don't move around as much. This can lead them to form interconnections between them creating crystals and other solids. At that point, things in the garage will never leave unless you burn the garage down, thereby heating them back up.

The laws of thermodynamics affect our lives in so many ways that we mostly just ignore. Thinking it through can get very complicated. For example, if the globe gets warmer, that condition might lead to more stuff in the garage. More stuff is necessarily disorderly. If, on the other hand, the climate gets colder, it may solidify the stuff in the garage, and we'll never get rid of it.

Here is an even more complex example. When a jet fly's high overhead shuttling politicians from here to there, the heat from the engine cools as it leaves the jet in the cold temperatures at that altitude. The water vapor in the exhaust then forms ice crystals creating the wispy, little contrails left behind the jet. As the little ice crystals fall and warm up, they move apart, through entropy, to form broader and thinner trails.

At the same time, the sun is warming the earth below so that things are moving further and further apart. This heat, we are told, is being trapped by greenhouse gasses produced by such things as jet engines. Both dynamics are happening simultaneously although at first the two events seem disconnected.

However, in the week after 9/11, there was a total ban on air travel so there were no contrails. It turns out, looking back on weather data, that the temperature of the United States went up one degree Celsius during that time. In other words, the jet contrails have a cooling effect on the earth by dissipating the incoming sun energy. There are about 100,000 commercial air flights a day, not counting military and private flights. So collectively, the contrails have a cooling effect on the planet.

This is yet another example of entropy. If you have an empty sky, something will fill it, namely airplanes, and the exhaust will freeze there. So, the jets cause global warming by producing greenhouse gases while cooling the earth with contrails.

It's a fine balance to be sure! I am certain someone has calculated the exact number of flights against the released carbon dioxide, against the amount of carbon dioxide required for plants to manufacture the necessary amount of oxygen to supply the expanding population which will soon collapse due to low fertility rates anyway. You'd think that kind of information could be made public to help citizens manage their garages more efficiently.

6 - WORK

I've tried to explain to my wife that t isn't as easy being a professional windbag as it looks. I spent years in preparation: undergraduate degree, graduate school, and scientific conferences. Somehow, she still thinks what I do isn't work.

According to physicists, work is equal to force times distance (W=F x D). Like me, you probably remember that from high school. It didn't make sense then either. So, does it mean if nothing moves, no work is being done? This is where the average person begins to lose faith in science. I have spent years building powerful muscles that enable me to sit for long periods of time without tiring, and now you're telling me it was all for naught?!

Because I sit unmoving at a computer most of the day writing words or music, she thinks that nothing is moving. So, if no force is being applied, no work is being done. See how misleading physicists can be. This is one reason they are seldom elected president or anything. You can't trust them to make sense. . .

What are all these mysterious forces anyway? Have you ever seen one? They tell me forces are measured in Newtons. But Newtons are simply an abstract idea arrived at by multiplying mass times distance divided by the time squared. You probably remember that from High School too.

$$N = \frac{M \times D}{t^2}$$

Anyway, in an effort to justify calling what I do work, I decided to investigate the process using physics itself to defend my activity. My hypothesis is as follows: if computer keys require a force to press down over a distance, then work is involved.

The average computer keyboard weighs 900 grams (gms). Since there are 106 keys on most keyboards, I can estimate that each key weigh about eight grams. Of course, that is an overestimate because keys are just the façade placed over the working mechanisms and contained within the surrounding frame. Therefore, I am going to estimate that each key only weighs about five grams or 0.005 kg. I think it is less than that, but I am fudging the data to support my hypothesis.

When a computer key is depressed, it moves an average of two millimeters (mm) or 0.002 meters. Since force is applied over time, I estimate that a key stroke takes about 0.2 seconds (sec.). So, I now have mass, distance and time with which to calculate the amount of work done in one key stroke.

$$N = \frac{M \times D}{t^2} \qquad \frac{0.005 \text{ kg} \times 0.002 \text{ m}}{0.04 \text{ sec}} = 0.00025 \text{ Newtons}$$

W = F x D   0.00025 x 0.002 m = 0.0000005 Newton-meters (= joule)

I admit that 0.0000005 Newton-meters doesn't sound like much. However, in a 650-word column, there are approximately seventy keystrokes per line and on average sixty lines. That is 3600 keys strokes which would be 0.018 Joules. Now even that is a low estimate because I only use two fingers to type which means I must move my hands a lot more than most people. I figure that fact alone at least doubles the amount of work to 0.0036 joules per column. That sounds like work to me!

Considering that I do this day in and day out for hours on end, I figure I do plenty of work. For example, my most recent book, "A Convenient Truce: a cease fire between religion and science", which is now available from Amazon, in both paperback and kindle formats, has 60,000 words for approximately 180,000 key strokes, or 0.09 joules. Ha!

Being a professional windbag is just as difficult as any other activity if you're really going to do it well. Some mass still must move over some distance within some time frame squared. What is squared time, by the way? Just because I make it look easy doesn't mean that it isn't all just hot air after all!

7 - PRESSURE

Daniel Bernoulli once said "½pu²+P = constant". This was back in the 1700's, and they talked funny in those days. Today they just call the equation Bernoulli's Principle. I'd tell you what all the little letters and stuff mean except that he was Swiss and I don't know that language either. However, I have been assured that he, indeed, did say it and that it is highly significant.

There are only a handful of ways of determining what one is to believe in this world. One of the most common methods is simply believing what we've been told by someone in authority whom we trust. That is why I think the earth revolves around the sun. (I could repeat the calculations and experiments of others. But I am told it is tedious, time consuming, and would interfere with playing the guitar.)

This is one of the reasons why college degrees and other acronyms are so popular to put behind people's names. It makes them look like they actually know something, and you should trust them. However, when one's own sister tells a younger brother that the sky is red, before he has established mastery of his colors, it becomes hard for him to ever really trust again.

Anyway, I believe Africa is there because everyone seems to think it is. I've never seen it. I also believe there are such things as atoms because I have been told that they exist. Accepting the existence of atoms allows us to tell interesting stories about how the world works. Of course, stories are not really proof of anything. Believing in atoms does save me a lot of time to play the guitar though.

Science and math have many uses. One of the main use of mathematics is to cover up human mistakes. The other is to confuse people. This is exactly what Bernoulli did. Prior to the 1700's people thought that the faster and harder something moved, especially liquids, the more pressure it exerted. (This was a huge mistake as demonstrated today by the numerous blowhards in public office.)

It is easy to understand how simple people from earlier generations, who were so ignorant compared to us modern folks, might get confused. I mean, if you've ever tried to wade a creek at high water like they did in the olden days, you'd think the water was pushing on you pretty hard. So, it took some mathematics to cover up that mistake.

Bernoulli discovered that the faster a fluid, or gas, travels across a surface, the less pressure that they exert on the surface. See? Doesn't that equation at the beginning make perfect sense now? Yeah, I know. Me neither.

Anyway, I'm sure Bernoulli is an expert although he didn't have a Ph.D. after his name. However, just in case you'd like to verify that the creek is sucking your legs out from under you, instead of pushing you over, try this little experiment.

Fold a sheet of paper in half like a little tent. Make sure it isn't your wife's special scrapbook paper. Just make sure, that's all I'm saying! Trust me. Now, set the tent on a table while you gently hold the sides of the tent to keep it from blowing away. Inhale and blow gently through the tent. Observe the sides of the tent as you blow.

When you blew through the tent, you were creating an area of fast-moving air. According to Bernoulli, the faster air travels across a surface, the less pressure it puts on that surface. By blowing, you lowered the air pressure inside the tent. The air pressure outside the tent is greater than the pressure inside, so it causes the sides to collapse.

In conclusion, just let me say $p + q = p_o$. The other stuff you can just ignore. You know it's true because Bernoulli said so.

8 - GEOMETRY

I was unaware that April was Math Awareness Month. My unawareness is either a terrible missed opportunity to publicize math awareness, or a great stroke of luck that I avoided further exposure of my ignorance. I think the reason math needs an awareness month is that most of modern math is disguised as quantity.

I guess quantity is pretty good as far as it goes. It's just that when it comes to the things most of us really want to know, we are generally more interested in quality. Like when I was trying to impress my wife with how much affection I felt for her prior to marriage, it would have been extremely foolish of me to say, "I love you about six."
Yet looking back, that is what I would have had to say if I expected to love her even more after 48 years together. I suspect we are all the way up to oh, say, eight or nine depending on how much longer I expect to live. However, if I were to proclaim any quality today, I would be forced to say ten. Otherwise I would not expect to live that much longer. See? Numbers have their limitations.

Take the number two. Pythagoras (And who am I to argue with him?) thought two was the first number that had sex. You know, two represents male and female, and all sorts of other opposites. Now right there, numbers get more interesting than just quantity. I think listing all the things that have opposites would be far more interesting and useful than jumping directly into the fact that two follows one, or that two plus two equals four.

Speaking of adding. Did you know the Greeks thought of numbers as squares or triangles? If you add one pebble plus two pebbles you get three pebbles which can be stacked into a triangle. If you add three plus three you get six which can also be a triangle of pebbles. Numbers like three, six, ten, and fifteen are triangle numbers.
However, if you add one pebble plus three pebbles it makes a square of four pebbles. Numbers like four, nine, sixteen and twenty-five all make squares. However, if you add any two

consecutive triangle numbers you always get a square number. That is why, to this very day, we think mathematics is for squares. And that's the reason why I like geometry. Geometry has personality. Yes. Square is a personality. With geometry, we can almost always tell when something is a good square or a bad one. (Down boy! Sit! Bad Square!) There are many ways a triangle can be a good one: equilateral, isosceles, or scalene. One doesn't confuse parallel and perpendicular. My wife says I'm not a square, but I do seem to be perpendicular. I'm not sure what she means by that, yet it seems clear. That's what numbers with personality can do for you.

I didn't learn about symbolic numbers until I sneaked a book out of the library and read it under the covers late at night. I hid it during the day under my mattress, although I'm not sure why. My mother was always afraid to come into my room. Yet, somehow, I sensed that I was on forbidden terrain.

Within its pages I discovered that bee cells, snowflakes, and quartz crystals all have the same six-sided shape. The reason is that six-sided objects are the most economical use of space. There is also a reason plates, pipes and planets are round. Hurricanes and hair curls both unfold as spirals. Coincidence? I don't think so.

The numbers from one to ten are all represented by a geometric shape. Their designs are found throughout the universe. By treating math as only quantity and book keeping, we miss out on a lot of more interesting personality traits. Quantity pales in comparison to the heptagon, symbol of the sacred virgin.

9 - TOPOLOGY

Reality can be sort of strange sometimes. Have you ever stared at something until it starts to look like something else? I mean, when you haven't been drinking. I haven't, but I am told some people can. It must be true because there are numerous people who can look at something and see it completely differently from the way I do.

If you're really good at seeing things differently you might consider a career in topology. Or, go see a mental health specialist. It's about the same thing. Topology is a field of mathematics that studies geometric problems that depend, not on the exact shape of the object, but on the way the object is put together. In other words, topologists study the reality of something not based on its real shape, but on the shape it might be if reality were different.

Yet another way of explaining it is that topologists study the properties that are preserved in an object when it is deformed, twisted, or stretched. Tearing, however, is not allowed. Deforming an object is not really changing the material, whereas tearing it is making a substantial change. I don't know, that's just what topologists say.

In topology, a circle is considered equivalent to an ellipse because a circle can be deformed into an ellipse by stretching. This sort of turns the whole idea of equivalence on its head. A sphere is topologically equivalent to an ellipsoid. Well, I don't know about you, but I just flat disagree with that. I don't know who makes up these rules. Well, actually I do know. Topologists make up the rules. As usual with mathematicians, I don't know what they are talking about.

Topologists do have some useful ideas though. For example, a topologist might have a hard time determining a doughnut from a coffee cup. See, if one were to twist and poke a pliable doughnut into the right shape, the hole could become the handle and one could poke a cavity to become the cup portion. It's the same structure, just twisted differently. Of course, it wouldn't really hold coffee.

Now before you get all really excited, I have already patented the idea. This is going to be huge - a doughnut with the coffee poured into it! No more messy dipping. We'll start with a simple glazed doughnut, but there should be no reason we can't expand quickly into cinnamon and cherry jelly.

Anyway, that gives you a feel for how important topology can be in solving scientific problems. Another major contribution of topology was the famous "hairy ball theorem". This theorem was first stated by Henri Poincare, a French mathematician in the 19th century. He stated that, "there is no non-vanishing continuous tangent vector field on even dimensional n-spheres."

That's what I'm told anyway. I think it's French. Translated, I think it means that if you try and comb a hairy ball flat, there will always be a cowlick. It took years of experimentation with comb-overs to prove the validity of this now accepted theorem. Cowlicks are real.

I had a topological experience once, and now I have a greater appreciation for this field of endeavor. It turns out that when a retina becomes detached (the retina is the field of nerve endings in your eye) someone either must solder it back in place with lasers or put a rubber band around your eye ball to push it back to the retina. It's sort of a "If the mountain won't go to Mohammad, Mohammad will go to the mountain" kind of situation. Wait, can I say that kind of thing?

Anyway, I ended up with a cool rubber band that distorts my left eyeball so that the eyeball itself is pushed back to the retina. Now, when I close my right eye and look at a round clock on the wall, it looks oblong. This has led me to question a lot of what I think about reality.

Apparently, one's brain can somehow correct for these kinds of things because now the clock looks round again. I'm back to normal, but it's not as entertaining as when I could switch back and forth looking at the clock with each eye. And I am still not exactly sure which version is real and which version of the world is simply a mental construct.

This raises the interesting question about whether topology is even real or not. I mean, if we stare at something and deform it in our minds, will our minds just un-deform it after a while? Perhaps what we see is reality that has been deformed by our brains and isn't really real at all. This may be a hopeful thought. Perhaps someday, the brains of all those people who see things differently from the way I see them will eventually un-deform and they will finally begin to agree with me.

10 - HANGING OUT WITH GRAVITY

I haven't always paid close attention to gravity. It's kind of a mundane subject. Gravity just always seems to be hanging around, and I take it for granted. How does it do that anyway? You'd think it would fall.

What is gravity, Really? What is kind of surprising to me is how long it took for scientists to discover gravity. I mean, Isaac Newton didn't discover gravity until the late 1600's. Why did people think things fell for the few thousand years before that? It appears that people just thought that "down" was part of the nature of apples and didn't attribute it to anything else at all. That seems like a strange view of reality to us today.

In fact, if you didn't know about gravity, why would one even expect things to fall down? Why couldn't an apple just fall up? Or sideways? Why did we decide to call the direction an apple falls "down" anyway? Is falling down worse than falling up? "Help, I've fallen up and I can't get down." Doesn't seem to make sense. I fell in love once and haven't ever been able to get out. How can we fall both in and out of love? Obviously, the rules of gravity do not apply.

I admit that down is significant, if for no other reason than that is where we fall. The act of falling can be quite invigorating if one doesn't think too much about the act of stopping the fall. I suppose even stopping falling is a scientifically fascinating concept in itself. The REAL technical term, I believe, is "ouch".

Newton actually did get his ideas from an apple although, apparently, the apple didn't hit him on the head. He told many people that he was out wandering through a garden when he saw an apple fall. It apparently just struck him one day, the idea, not the apple, that something had to be pulling the apple towards the earth.

It has always made me wonder a little about Newton. I mean, didn't he have a job to do or something? Who has time to wander about in gardens? Was he watching apples fall, or did he just sort of see it happen out of the corner of his eye? Anyway, there are two great lessons to be learned from his experience besides the fact that Newton was a strange fellow.

First, this story demonstrates the value of wandering around outdoors in gardens instead of going to school. If he'd been in class he would never have discovered anything. It also shows how science is first and foremost a matter of asking strange questions.

It's not as if people hadn't already had rich experiences with gravity from just walking around, tripping, swinging from vines in the trees, and falling out of bed. There were plenty of observations, I suppose. What was needed was for someone to ask, "Why do things all fall in the same direction?"

This is the second lesson. His answer was to declare that there must be some sort of invisible force that acted over distance to attract things to it. At the time, some people accused him of dabbling in the occult. I mean, really? An invisible force? You must admit it sounds suspicious.

His real contribution though was to ask, "How high does gravity go?" Newton decided it had to go quite a way into the air because the apple tree was pretty tall. This led him to speculate that perhaps this invisible force extended all the way to the moon! Even he was surprised to calculate that it apparently went all the way to "infinity and beyond", to quote Buzz Lightyear. In fact, his theory and calculations settled the question of how and why the moon circles the earth and the planets circle the sun.

Anyway, Newton did not receive any compensation for his most famous slogan: "A body at rest tends to remain at rest unless an external force is applied to it." That is only one of his three wise sayings. Knowing this makes you now one third as smart as Isaac Newton.

Unfortunately, he never explained how to avoid those external forces. I am working on that problem by spending as much time as possible remaining at rest in the garden. You might say that gravity has kind of got me down.

11 - RUN DON'T WALK

When I started my scientific career, my major advisor gave me some advice. I mean, what else are advisors for? He certainly didn't do any of the work! One thing he said was that there wasn't much of a future in Parasitology because we had made so many strides in treating and preventing disease.

Bad advice. Parasites are still doing just fine.

However, one of his better pieces of advice was to stay away from doing research on Malaria. He explained that there has been so much research on this disease that it would take me a life time just to catch up on what's already been done. Then, if I discovered anything new, I would probably just be one of fifteen authors of the paper.

He suggested working in obscure fields where discoveries are easier to make, and the credit would be all mine. I believe I have succeeded beyond his wildest dreams. I probably couldn't be more obscure, and the credit is all mine.

Having received tenure, I am now ready to boldly tackle some of the major problems in science. There is one problem that has intrigued me since childhood when I used to have to walk everywhere I went. (This was in the olden days when children could do that.) Is it better to walk or run when it rains? If you walk, you spend more time in the rain. If you run, you are hitting some rain drops from the side adding to the wetting potential.

If you stand still, the only rain drops than can hit you are the ones directly over your head. The surface area exposed to rain of an upright parallelogram is not dependent on its height, just its depth and width. Therefore, all the rain drops directly over your head will strike you whether or not they fall, or you rise upward into the rain. This would be one more disadvantage of having a big head.

If you walk, or run, the number of rain drops hitting you from above stays the same from one position to the next, so the rain from above is a constant value. How wet you get then depends simply on how long you stand in the rain.

If you wish to get out of the rain, you will have to move towards shelter. In that case you will also run into rain drops from the side. So obviously, you will stay driest if you don't move at all. Well, I guess that depends on how long it rains and how long you stand there in it.

The number of rain drops you will run into from the side is dependent on how far you must travel to get to shelter. There are only so many raindrops between you and your destination. Since this is a function of distance it will be the same no matter how fast you might move.

So, the total number of raindrops that will hit you will be the number of raindrops falling per second from directly over your position at each second, multiplied by the number of seconds in the rain, plus the number of raindrops per meter hitting you from the side, times the number of meters it is to shelter. It might look like this:

Total drops = drops per second x # sec. + drops per meter x # meters

You can't change the number of raindrops falling per second or per meter and you can't change the distance to shelter. Since the number of drops from above depends on time and the drops from the side depend on distance, and since the distance is set, the only way to reduce wetness is to reduce the time spent in the rain.

So, like a good advisor, my advice is to run. However, as usual, more research is needed.

SERIOUS BIOLOGICAL SCIENCES

It isn't totally clear what physics and chemistry have to do with biology. A lot of physical scientists claim that being alive is just the sum of all the chemical reactions that occur in a living thing. They believe we are all just machines and that even our thoughts are just the sum of the chemical reactions in our brains.

I am not sure I believe this because then the idea that thinking is just an elaborate machine came from someone's thinking with their elaborate machine. Can I really trust someone else's machine to create significant ideas? So, all the mathematics used to manipulate the physical world are the result of a mechanical brain that created the mechanical ideas to manipulate the mechanical world. I think it's a house of mirrors.

Of course, those who believe that we are just machines seem to think that they are superior machines and their ideas are therefore the result of something more than a machine. Or their machine is better than mine. But I think I am getting confused here so let me get back to biology.

Living things seem to fly in the face of the laws of thermodynamics, promulgated by physical scientists without any actual proof. This is because living things become more organized and complicated rather than falling apart according to the concept of increasing entropy.

Of course, the physical scientists claim this is because energy is added to the system. Well, I admit that energy is added to the system, but did you know that it has never been proven that the energy added is the sum of the energy required. Careful measurements of energy in and growth have seldom been made, and when they have the results are wildly inconsistent.

Of course, the supposed central tenant of biology is evolution. This is a gross mistake that has wasted countless years of understanding. Mine is distinctly the minority view of course. But consider this. Evolutions basic concepts are:
- that all living things produce more offspring that is required to sustain the species.

- The environment selects the progeny who will survive to the age of reproduction.
- Only those most capable of survival will be allowed to reproduce, thereby changing the nature of the next generation.

Well, that's cool as far as it goes. However, the new generation of offspring, being supposedly more efficient in some way than the last, will itself change the environment. So, the very feature that selects survival changes with each generation. It's a dog chasing its tail. Therefore, evolution can be used to examine the past in some way, but can never be predicted or totally controlled.

There are a lot more interesting and significant ways that a materialistic model isn't totally useful in describing the living world. I will discuss more about the living world in chapters twelve through seventeen.

12 - ORGANIC

I am so inconsistent it is reliable. I can be self-disciplined when I want to be, but not at all if it isn't my idea. I guess that makes sense. If it isn't my idea, it isn't self-discipline. If it's someone else's idea then it is other-discipline. I hate that. Still, sometimes I just can't depend on myself. Isn't that what wives are for?

However, with age, wisdom, and greater skill in self-justification, I have decided that I am not a flake. I'm just organic. Organic, contrary to popular opinion, doesn't mean raised naturally or without chemicals. That's just marketing talk. Everything is raised with chemicals. The difference between organic and nonorganic is merely a matter of where the chemicals come from. Do you like your nitrogen coming from mines, petroleum, or manure? If you prefer manure, you are organic.

In science, the word "organic" refers to things made of carbon. Carbon is a cool atom because it is relatively abundant and can bond to other atoms in multiple ways. This allows it to form long chains and complex shapes such as rings and spirals. I have fresh appreciation for organic when I realize that my good-looking wife is mostly just carbon and water tension.

Carbon may not sound very exciting. But it gives organic things, like me, a great deal of flexibility and strength which inorganic substances sometimes lack. Most organic, living things can seem contradictory or inconsistent at times. Except my wife, of course.

For example, trees can be strong and flexible. That's why wood makes such good medium for musical instruments. It can be strong enough to withstand high string tension yet pliant enough to vibrate with the strings. Sponges are soft but have rigid spicules within. That's why natural sponges are good for cleaning surfaces. The tiny, hard spicules imbedded in the tissue make good abrasives.

Manmade objects have very different characteristics. I think this is because humans tend to focus, when they aren't being inconsistent, on a specific characteristic required for a task. Because humans are inconsistent, we go to great pains to consistently concentrate, design, and build things for specific and narrow purposes. At least it seems like bosses do.

We make one kind of steel that is strong and inflexible, and another kind that is flexible but not as strong. We do the same thing when we design homes or communities. We design neat places in which to work, but design other places to live in that are usually far away from our work. We have living rooms that are seldom lived in because we live in the family room. The living room isn't for the family.

Humans have an extended learning period to function in the world. So, we place them in unnatural worlds far from their homes, for most of every day, to learn. Humans have stores to buy food, but we put our farms often thousands of miles away.

Humans create drainage districts to promote water runoff. Then they spend money on equipment and management practices to improve water-holding capacity of the soil. Interestingly, good top soil promotes both drainage and water-holding capacity. I suspect the average American rarely thinks about top soil.

I have discovered the value of inconsistency. It encourages one to discover that things can be done other ways. Does our world really have to be the way it is? Could people live in small communities scattered across the land where they could live close by and educate their own children? Does food have to be shipped, on average, fifteen hundred miles to be consumed? Could I be both rich and good looking? Wouldn't you know, the thing I am consistently good at is neither.

Sometimes, when I am particularly organic, I discover I like the new ways better and try to become consistent in those new ways. It never works. I just move from one inconstancy to another. You'd think that, with six thousand years of recorded history, humans would have learned the value of manure, I mean organic.

13 - HOMEOSTSIS

Have you ever wondered why cows produce "cow pies", horses produce "road apples", and sheep produce "sheep pellets"? Why don't we call sheep droppings "sheep grapes" or "mountain grapes", or some other kind of fruit? It's inconsistent. I don't mean to be crass or anything. It just seems odd.

Then there is the question about why three grazing animals produce such differently formed waste products. I know horses aren't ruminants, but sheep and cattle are. They all eat basically the same things but their waste is shaped differently. It's weird.

People ask me where scientists get their ideas. Personally, I always waited until they went on sale at the dollar store, but all scientists probably have their own method. It used to be that people got ideas from the natural world around them, but today's scientists have limited experience with the real world. Mostly they investigate previous discoveries made in the lab.

For example, never having been shepherds, most modern biologists would never wonder about the distribution of parasite eggs within the "mountain grapes" of sheep. I mean, are all the eggs on the surface, clustered in the center, or embedded randomly within the pellet? Regardless of the answer to that question, how and why are they where they are?

Don't try to answer this question by cutting sheep pellets into microscopic slices for examination using your major professors' microtome. The packed cellulose in the pellet can destroy a microtome blade and endanger your degree status.

I bet very few people today know that rabbits, which are generally herbivores, produce a cloudy urine that is basic in pH. I suppose my granddaughter knows this as she raises rabbits. But aside from a few rabbit fanciers and a couple of veterinarians, who knew? I realize one might well ask, "Who cares?"

Okay, the truth is I didn't know rabbit urine is supposed

to be cloudy and basic either until I read about Claude Bernard who did know this. Claude was a French physiologist in the first half of the 20th century. He is the one who knew this little tidbit about rabbit urine. So, he was intrigued when some rabbits, brought into the lab from a local market, proceeded to urinate on his table with a clear and acidic urine.

He also happened to know that most carnivores have clear and acidic urine, so he wondered if the rabbits had been eating hamburgers or something. He thought it more likely that they had not been fed recently, and were in an advanced state of fasting. It dawned on him that a fasting animal might be digesting its own tissues, thus causing t to metabolize as if it were a carnivore. First, by feeding the rabbits a normal diet he observed a change to cloudy and basic urine. Then feeding the rabbits a meat diet, by trickery, he once again observed clear acidic urine, verifying his theory.

Somehow this discovery led him to propose the idea of the "milieu intérieur", now known as homecstasis. This is the idea that variables in the body are regulated so that internal conditions remain stable and relatively constant in the healthy individual. So basically, Claude is the guy who started making people pee in a cup when they go to the doctor. The doctor needs to know if he is dealing with an herbivore or a carnivore, I guess. I know I would treat the two kinds of patients differently.

Well, that is an example of someone performing experiments based upon informed observations of the natural world. I'm wondering what promise that holds for a new generation of scientists who have never caught a crawdad or been urinated on by a freshly-caught frog? I can tell you that frog urine is clear, although I have never checked its acidity. My wife thinks this column is in poor taste. But you must admit the whole thing is odd.

14 - STRUCTURE AND FUNCTION

If structure does dictate function, which is one of the central mandates in the study of Anatomy and Physiology, then my function must be to mostly sit around. I have tried explaining this to my wife and, despite her Master's Degree in Science Education, she doesn't seem to understand. She somehow thinks that it is function that determines structure. She says that if I sat around less, I would develop a different structure.

Analyzing humans, as if they were mechanisms, has similar shortcomings. For example, when we perceive living things as machines, we treat them like machines. However, if we treat them like machines, we then perceive them as machines. Because science sees the world as mostly material, this perception may be one drawback.

Ironically, it wasn't a scientist who first made this observation. It was poet William Blake who observed, "What seems to be, is, to those to whom it seems to be." I don't think this thought has been properly appreciated, probably because of his use of the word "whom" which confuses people to this day.

Structure can sometimes provide interesting insights into human behavior, though. For example, the body is double-wired with information cables. One set of wires receives information from outside the body and is taken to the brain. The other set goes out, away from the brain, and carries response information to the body. Messages only travel in one direction.

Actions are mostly determined by incoming information because most of our actions are in response to incoming information. Consider that a crying newborn is not "thinking" it wants to cry. Crying is the only action it can control in response to the stresses of being born. Most early responses, on the part of babies, comes in response to stimuli it receives.

Obviously, this gets more complicated as the individual develops control over the body as learned actions and social norms have a greater influence on behavior. By the time we are several years old, it may seem to us that our behaviors are initiated internally. Indeed, as our brains increase in complexity, some behaviors may be initiated internally. But by then it is unclear whether the thought to initiate something doesn't come because of previous input that ingrained the behavior.

These ideas have serious implications for learning. When one studies by incoming wires only, seeing and hearing, they may understand an idea but be unable to perform any related action because they have not yet trained the outgoing wires to respond. Knowledge without skil is not only useless, but it is boring as it engages only half of the central nervous system.

Students recognize this though most are unable put the thought into words. Extracurricular activities are popular because they are some of the few things students get to learn and DO. The ability to be able to do something partially explains the fascination with cell phones and computer games.

One can learn to read a foreign language more easily than to speak it because speaking involves doing something with muscles of the mouth. It is easier to learn to speak a language than to write it because writing requires the more careful use of fine-muscle control.

On the other hand, it is almost impossible to learn to do something without learning "about" the activity. Even people who learn to do something poorly recognize the process and the abstract ideas stemming from their actions. The structure of the nervous system dictates that to learn something; we must do something.

Conversely, when you do something, you learn something. Assuming that "learning is a change in the structure of the brain," we can conclude that, as my wife says, the function also dictates structure. So, I must concur with my wife that function also dictates structure. All I can say, at this point, is that this is a most unfortunate conclusion to this column.

15 - BOTANY

Some people, mostly botanists, have claimed that I don't write enough about botany. They are probably right. So today, I will address the imbalance. But be warned. Everyone knows that violence sells newspapers, so I have warned the editor to print extra copies of this week's Saturday run.

There is a double standard used by many scientists to curry favor and research grants. For example, physicists have portrayed themselves as great beneficiaries of mankind, all the while providing us with the atomic bomb and electricity. Did you know that almost 4000 people a year are killed by electricity?

Chemists also like to think of themselves as benefactors of mankind because they can make plastic and pesticides. Now we are drowning in both.

For some reason, the zoologists get pegged with all the dangerous animals from lions, tigers, and bears, to rattle snakes, scorpions and wasps. Does anyone ever mention the bees and beetles that pollinate our crops, or the dung beetles that help decompose dung?

Then there are the poor, little, peace-loving botanists who just grow flowers, food, perfumes and hallucinatory drugs. They are not so innocent, let me tell you.

Take Aconite for example. Well, no, don't take it literally. Aconite is known as Wolfs bane or monkshood, and ingesting it lowers the blood pressure and stops the heart. The whole plant is highly toxic, but the root resembles a horseradish root or a pale carrot, so I guess vegetarians sometimes could get confused

Everyone covers up their wall sockets when children are born, but nearly 70,000 people a year are poisoned by plants. You think because you don't eat horseradish you're safe? Well, that Philodendron can cause nausea, severe abdominal pain, and serious allergic reactions. Dieffenbachia can inflame and paralyze the vocal chords leaving a person unable to speak. It can cause much worse symptoms too, so don't get any ideas. I could still write. . .

Sure, tell me about the kind, peaceful botanists that harbor such things as Castor Beans, the plant that gives us ricin. It is now obvious that my mother was trying to kill me when she gave me castor oil as a child. Supposedly it was a laxative. Oh sure! This is the same stuff the KGB used to kill Giorgi Markov, the BBC Journalist, back in 1978.

Okay you vegetarians, what about Lathyrus sativus, or Grass Pea. Sure, it is a good source of protein, but it also contains beta-N-oxalyl-diamino propionic acid. You know, that's the neurotoxin that kills the nerve cells in the lower limbs and leaves a person paralyzed from the waist down. They say that if you soak it long enough in water, it is safe to eat. How long?!

It isn't an accident that Alfred Lord Tennyson wrote of the Yew tree, Taxus baccata,

"The fibers net the dreamless head
Thy roots are wrapped about the bones."

In fact, in 1990 a yew tree was uprooted during a storm, and skeletons were found in its roots.

The whole plant is poisonous, but the sweet tasting berry causes the biggest problem. It contains a poisonous seed, and eating the seed causes a drop-in pulse rate and heart failure. Scientists seem a little unclear about the exact symptoms because often the first evidence of yew toxicosis is apparently death.

Don't misunderstand me. I don't blame the botanists that these plants exist. They didn't create them. It just gets a little tiresome hearing all about the healthy plants, garden clubs, and recipes for innocent Habanero Chili, when there is a dark side that is hardly ever acknowledged.

I have nothing personal against botanists. Some of my best friends are botanists. I think botanists should have the civil liberties accorded to any minority group and should never be persecuted. I think it's kind of like a birth defect or something. But please, just stop pretending to be all more innocent and innocuous than everyone else.

16 - GREEN BANANAS

I can't decide if I should tell my wife to stop buying green bananas or not. On the one hand, I am of an age where it could be tremendous waste of money. On the other hand, green bananas can be a source of youthful stimulation. It's a curmudgeon conundrum. (Technically a curmudgeon is any ill-tempered person, which, by the way, I am not at all. However, there is an accepted convention that they are also quite old, which, by the way, I am.)

I think green bananas are only a stimulant in plants. This is because green bananas are a great source of ethylene, a gaseous plant hormone that has several different functions depending on the plant and the stage in the plants life cycle. In some cases, it stimulates the release of dormancy, which, by the way, I could use. It also influences such plant activities as root and shoot growth and fruit ripening.

However, ethylene, the plant hormone, is not to be confused with ethylene, the hydrocarbon. Well, except for the fact that they are the same thing. You can tell them apart though because the hydrocarbon is a colorless, odorless, and flammable gas whereas ethylene is a colorless, odorless, flammable plant hormone. Anyway, I think the plant hormone is flammable. I've never actually tried to burn bananas.

I am told that both molecules look like a big X, with two carbon atoms double bonded in the middle and two hydrogen atoms attached to each carbon. That's the beauty of chemistry. Since no one has ever seen any of these molecules you can claim that they look like whatever you want and your idea is as good as the next guys.

We did do an interesting experiment with ethylene once. There really wasn't any point to the experiment except to make fun of a friend on his birthday. Clarence was always making a big deal of his own birthday, reminding folks when it was months in advance so we wouldn't miss it. So, one year we made a hall display of a bunch of green bananas under some sealed bell jar two months in advance of his birthday.

The bananas ripened right on cue because green bananas give off ethylene. However, they never spoil when sealed from outside air. So, for two months we watched the same bananas looking yellow and beautiful under the bell jar. On Clarence's birthday, we had a party and broke the seal of the bell jar. The bananas turned black and into a pool of banana smelling mush right before our eyes during just a few minutes. It was the best kind of experiment: quite disgusting.

Clarence pretended to be offended at the implied metaphor for his age. But secretly I think he was pleased that we remembered his birthday. To this day students recall this educational experience. . . I think. You begin to see how much fun science can be.

I never did find out why the bananas didn't rot when sealed. I did a little reading and learned that ethylene production from bananas sealed in a jar can be quite variable. I lost interest after a while though because doing research on a colorless, odorless gas in a plant that never moves is not very stimulating. Oddly, there are some peculiar people who find plants interesting. Now, I find that interesting.

So, you see my conundrum here. Just like other stimulants like coffee, once you are exposed to too much you become dependent on it. Once it is removed you melt into a big pile of black mush. So, I am not sure that having green bananas around the house is a very good idea. It might help preserve my youthful appearances if I were sealed in with them. But if we were to run out of bananas things could get very messy fast.

17 - SKUNKS

Neil deGrasse Tyson once said, "If you are scientifically literate, the world looks very different." I don't know who the heck he is, or how he knows, but I agree with him. You probably don't think you are very interested in a chemical compound called thiols. In fact, you probably think you aren't very interested in chemistry at all. That's alright. Hardly anyone is. However, sometimes it has its uses, especially around this time of year.

Chemistry becomes pertinent in the summer time for several reasons. It's chemistry that causes explosions, which everyone used to like, but which now seem to make people nervous. When I was in elementary school, Neil Stewarts Dad used to help us blast tin cans in the air with gun powder in a vacant lot nearby. Social services would have him arrested today.

I think I learned more in vacant lots than I ever learned in school. I wonder why they don't hold school in vacant lots. It's also chemistry that makes the bright-colored fireworks that we never get to watch anymore because of Al Gore.

Chemistry also provides us with sun screen. That's the magical ointment we put on to protect us from getting cancer, so we can develop Vitamin D deficiency instead.

These are not the real reasons why chemistry is important in July though, or why you should develop an interest in thiol. Not all summer hazards can be neatly corralled, isolated, fenced in, prevented, or avoided. Some literally come seeking trouble.

> "Crossing the highway late last night . . .
> He should have looked left. He should have looked right.
> He didn't see the station wagon car
> The skunk got squashed, so there you are.
> You got your dead skunk in the middle of the road."

I'm sure you recognize these famous words to that popular song by Louden Wainwright the III. Yes, it's skunk season! They are prowling all over the highways and byways of Colorado, getting smashed by inattentive drivers. A "Dead Skunk in the Middle of the Road" can be sensed upwards of a mile depending on wind conditions and temperature.

Thinking about that song does bring back fond memories of the time my four children performed the piece for our church social. Of course, my wife didn't know they were going to do it. Honest, I don't even know where they learned it. For those who don't know, it is a surprisingly tender song about lonely deaths on dark highways. It's not just skunks either. It also references dead dogs, frogs, rabbits and raccoons. It's a dangerous world out there.

But I digress. This is the part about your interest in chemistry. Skunks possess anal scent glands from which they can discharge a chemical composed of three low-molecular weight thiols which contain sulfur. These are detectable by the human nose at concentrations of only 10 parts per billion molecules.

What is worse is that thiols are insoluble in water which means they will not wash off skin, clothes, furniture, or your dog's fur, even with high powered detergents. This is where chemistry comes in handy. Dr. Paul Kerbaum published a solution (pun intended) in the October 1993 issue of Chemical and Engineering News.

He reasoned that the thiols must first be oxidized into water soluble sulfonates. Hydrogen peroxide is an oxidizing agent that will do this, but baking soda needs to be added to liberate the oxidant from the peroxide. The resultant sulfonates can then be removed by a small amount of hand soap.

The exact proportions are one quart of 3% hydrogen peroxide, one quarter cup of baking soda, and one teaspoon of hand soap. Unfortunately, if you try to make this in advance and store it in a bottle the thing would explode, which could be a nice, safe addition to the 4th of July celebration. But it might get you in trouble with social services. That would really stink!

SERIOUS HEALTH SCIENCES

Let's be honest here. I have no idea what health sciences are, and neither does anyone else. Health is an abstraction and is only vaguely defined. In fact, if you examine numerous definitions you will find that basically health is the absence of disease.

That sounds pretty good if one is clear about what disease means. No one is. If you examine numerous definitions of disease you will find that it basically means the absence of health.

See, I could just as easily call this section of the book Disease Sciences. I was tempted to do exactly that, but I realized that it would sound like I was promoting disease and it might be hard to get funding for that. That's why everyone has switched to health sciences. It's like being able to say that you are all for health. Much more positive.

A more accurate title would be something like Applied Biology, except there are a lot of ways of applying biology and not all of them are healthy. Poisoning, pollution, torture all come to mind. I have not included any chapters on these topics. Well except for the one on botany, but they deserve to be brought down a peg. Well, the one on how to not die could be twisted into something negative I guess. As for the chapters on metabolism and arthritis . . . I guess it's probably better not to give you any ideas.

Just take my word for it that these chapters, eighteen through twenty-three, taken collectively, and in the right frame of mind, will increase your health and wellbeing, or at least decrease your diseases.

18 - EXERCISING MEMORY

One of the big questions in science has been how to get rats to lift weights. The rats simply refuse to do bench presses. Until we could get rats to lift weights, we had no way of comparing aerobic exercise, like walking and running, to anaerobic exercise, like weight lifting.

I suppose we could have used humans, but they are so unreliable. Actually, scientists tried using elderly females in some experiments. (I hasten to make a disclaimer here. In spite of this fact, I will be making no snide remarks concerning gender in this column. My wife is my proofreader.) However, scientists found it much easier to extrapolate rat data to humans than from elderly females, so we were left with the problem of weight lifting for rats.

The process turned out to be surprisingly easy. The scientists simply tied weights to the rat's tail's and put their food at the top of a ladder. This method of using a heavy tail, along with food as incentive, has been a great benefit to me personally. It has enabled me to build powerful muscles that enable me to sit for long periods of time without tiring.

We needed rats to lift weights because we already knew that if rats had access to a running wheel, they produced more cells in the area of the brain controlling memory and performed better on memory tests. We didn't know if anaerobic exercise such as weight lifting would have the same effect.

Scientists at the University of Columbia began to think that weight training might improve memory when they did the previously mentioned experiment with elderly women experiencing early, mild dementia. The women were assigned to groups with prescribed exercise regimens. Some did mild weight training, others did stretching and toning, and the third group walked.

They determined memory loss by asking the participants to find their car keys and recall their husbands' names. The stretched and toned group failed miserably. The other two groups only failed in recalling their husband's names, indicating a clear benefit from both exercises.

Unfortunately, when these ladies were scheduled to come in for post mortem brain analysis, they all forgot.

So once again, we had to turn to rats. It turns out that it really didn't do the rats who were trained as runners any good. They still got caught and their brains analyzed. These "runner-rats" had increased levels of something called BDNF (brain derived neurotrophic factor). This compound apparently supports neurons and helps grow new neural extensions.

Weight trained rats didn't have more BDNF, but they did have an insulin-like growth protein that also promotes cell division and support. So apparently weight training, or in this case tail training, does have its benefit for memory also. Unfortunately, it's not clear which protein is used by elderly females in husband recognition and key finding.

More research is needed (isn't it always?) before these results can be applied to actual humans. I'm not sure I have a lot of faith in this data, however. My grandpa started walking five miles a day five years ago. But he forgot to come home, and now we don't know where he is.

Personally, I am not sure that having a better memory is worth the extra work. The longer I live, the more I mess up. There are a lot of things I'd just as soon forget. It seems like not being able to find the car keys would be a small price to pay, in aggravation, for the ability to put certain things out of my mind. However, we can all rest easier now knowing that one of the big questions in science has been answered: how to get rats to lift weights.

I want to thank Dr. Ed Bonan-Hamada from the Colorado Mesa University Math Department for sending me this information. I'm not exactly sure, though, why he thinks I needed it.

19 - ARTHRITIS

My thumb really hurts. I wrote my granddaughter a letter by longhand while sitting in a boring meeting. Now my thumb is killing me. I must have injured it earlier in the day or something. My wife thinks it might be arthritis, but that can't be because arthritis only happens to old people. No, I'm sure it's an injury. It will be better in a few days, or as soon as I take four Ibuprofen, whichever comes first.

There are a lot of people who suffer from arthritis though. I understand it is a painful condition that feels a lot like the joint has been injured. It's always a joint. In fact, the term "arth" means joint in Latin. Before I became a biologist, I thought a joint was a place like Dirty Charley's where I learned to play pool. I guess there are different kinds of joints. But I don't think you can smoke an arth.

The "itis" part of the word means inflammation. So, whenever you see "itis" stuck on a word, it means that the word is inflamed. Bronchitis means the bronchi are inflamed. Houseitis would mean your house is in flames, I guess. Come to think of it, an arth can be inflamed, so perhaps you can smoke them.

That leads me to what being inflamed means. One of the first things learned as a scientist is that nothing in science ever means the same thing to scientists that it does to real people. I mean a joint, which is called an arth, is never actually in flames when one has arthritis, unless it is Dirty Charley's which I believe did actually burn down some years ago.

Inflammation is pretty simple. Whenever your body is insulted the result is inflammation. It doesn't matter whether the event is a cut, a shoe rubbing on your heel, or a bacterium irritating the back of your throat, your body pretty much responds the same way: by dilating the blood vessels in the immediate area.

This seems like a pretty good idea. When the blood vessels expand around the injury, they bring in more blood to the area. This is what causes irritated tissue to become redder than normal. Since more blood means more oxygen and nutrients for the injured cells, you'd think this would be a positive thing. The extra blood could even serve to flush the area of contamination or toxins.

This extra blood comes from the core of the body where there is a larger volume of it present. But the core of the body is also hotter than the typical 98.6 F that we consider normal body temperature. That's why injured areas, like a sprained ankle, often feel hot to the touch. The injury looks and feels like it's on fire.

Of course, the extra blood also brings extra fluid into the area causing swelling. The swelling cause's pressure on adjacent neurons, and the result is pain. When something in my body hurts, like my thumb, I tend to stop using it. So, I call the doctor and tell him it hurts to use my thumb and he tells me to stop using my thumb. Of course, I had already stopped using my thumb because it hurt to use it. What I want is to be able to use my thumb again. We needed the Affordable Health Care Act for this?

All that sounds pretty good except that the swelling can become so great that it damages more tissue which leads to more inflammation. More inflammation causes more swelling, which causes more tissue damage. If this inflammation is happening in your "arths" it damages the cartilage that is supposed to cushion the two bones. This causes more inflammation. Soon the arth is so painful you need a prescription for a medical "arth" to fight the pain. It's a vicious cycle. But mines just an injury, I'm sure.

20 - WHISTLING DIXIE

I'm not so sure that whistling while you work is such a good idea. I used to whistle "Cherry Pink and Apple Blossom White" all the time in the biology lab. I liked the way the fourth note hung out there for as long as I wanted before I staggered it down. Then one day a delegation of graduate students held an intervention. They thought I whistled fine, but the inconsistency in the length of that fourth note was driving them crazy.

My sister taught me how to whistle while we were swinging on some park swings. That was way before I knew anything about sound vibrations and air waves. Now it seems strange to me that a seemingly smooth flow of air would generate oscillating sound waves.

People are like that, I think. Once we learn how to do something, we don't give much thought to how we learned it or how we do it anymore. Most of the time we can't remember when, where, and how we learn many things. When did you learn to read? I bet you don't even know how you swallow. (Note to self: future column on swallowing.)

Did you know that there are two holes on a standard whistle-kettle cap? It's true, and I had to look long and hard to find a tea kettle with an intact lid to verify that fact. The two holes are crucial to whistling according to recent research done at Cambridge by Ross Henrywood and Anurag Agarwal.

They explain that when the water in the kettle is heated to steam, it is pushed out of the kettle by internal heat pressure. This creates a stream of moisture laden air that escapes through the first hole. This stream quickly becomes unstable because, as soon as the pressure is released, the steam begins to coalesce back into droplets. It's like water from the nozzle of a hose goes from being a stream to being droplets again after traveling a short distance.

Then the unstable stream is immediately forced through the second hole. Because the stream of air is now unstable, it arrives at the second opening in an irregular series of pressure pulses which cause the stream to exit the second hole in a series of vortices, or spirals. These spirals of air pressure create sound pulses of air waves that are very close together. It's these that create the high-pitched whistle.

All that sounds perfectly plausible and interesting, if perhaps a little useless. I was doubtful of their theory at first because I didn't see how it explained "Cherry Pink and Apple Blossom White" as whistled with the traditional pucker-whistle humans use for melodic whistling. I therefore conducted a series of dangerous, and sometimes painful, experiments to discover the mechanisms behind pucker-whistling.

The pucker-whistle is produced by placing the tip of the tongue behind the front bottom teeth, puckering the lips, and blowing. This appears to be a single opening with no opportunity for the flow of air to be turned into pressure pulses. However, careful and painful analysis indicates that the tone is changed by altering the position of the back of the tongue. The back of the tongue forms a restricted opening through which the air must first pass before leaving the second opening created by the tongue and pursed lips.

So melodic whistling has some characteristics in common with tea kettle whistling after all. Why didn't they just say so? Now that's important!

Ironically, it was Hermann Ludwig Ferdinand von Helmholtz, an early pioneer in acoustic research, who once observed, "Whoever, in the pursuit of science, seeks after immediate practical utility may rest assured that he seeks in vain."

I, building on the shoulders of Henrywood and Agarwal, suggest that by altering the shape of the tongue, the fourth note of "Cherry Pink and Apple Blossom White" can be sustained for a variable duration which drives other people crazy.

21 - HOW TO NOT DIE

When something happens that we think is real, humans always ask, "Why did that happen?" Curiously, we're almost always interested in explaining negative events. We seem less interested in understanding how positive events happen.

For example, if a plane crashes, humans gather and analyze data to find out why. Well, duh! It fell because of gravity. The really amazing question is, how in heck did we get thousands of pounds of metal and people up in the sky in the first place?

When there is an auto accident, police investigate to determine the cause. I'm sorry. I must be missing something. Two people obviously tried to occupy the same space at the same time. If that happens slowly enough, it is just a shove. At higher speeds, it's called an accident . . . or an assault.

Why is there too much carbon dioxide? Maybe we should find a use for it. Oh yeah, that's called plants.

When people die, humans record the cause of death on the death certificate. Why don't we just say birth? One cannot cease to exist until after they have been conceived. Once we're born, death ultimately follows.

Another thing we agonize to death is poor people. Wouldn't it be better to ask how to get people rich, than how to help people cope with being poor?
Energy supposedly cannot be created or destroyed. Who said? So where did it come from? And where is it when I don't have any? It must have been created sometime, somehow. Am I supposed to just accept this on faith? I think it would be cool to learn how to create energy, not just move it around.

The causes of death are certainly interesting in a morbid sort of way. Personally, I am more interested in knowing how to stay alive. I guess the theory is that by examining causes of death we can better understand why people die. Well, besides just being born. I think birth was a fine idea in my case. Others may disagree. I just think it would be more useful to know how to not die.

They tell me that the number one cause of death is heart disease. Just like airplane crashes, that really tells me nothing. Heart disease merely tells us that the heart didn't work properly and quit. I think it is more important to tell people how to make their heart work better.

Interestingly, we already know how to make the heart work better. The single most important thing people can do to make their heart work well is to not smoke. It doesn't seem to matter much what is smoked: dope, regular tobacco, or seed pods. I smoked seed pods once as a kid and almost died of respiratory failure.

Of course, smoking is also related to lung cancer, chronic obstructive pulmonary disease, breast cancer, high blood pressure, pneumonia, and stroke. In fact, this list of diseases makes up five of the top six causes of death in the United States. Really, the best way to not die is to not breath in smoke.

If you are really serious about not dying, you might also consider avoiding accidents, violence, and liver disease. These three rounds out the list of the top nine causes of death in the US. However, these problems can be minimized by eliminating alcohol, marijuana, and illicit drugs from your indulgences.

We could save a lot of administrative costs if we were just a little more honest with our questions. Instead of asking why the plane fell from the sky, let's ask how to keep it up there. Instead of asking how to avoid accidents, maybe we should just always let the other fellow go first. Instead of asking the cause of death maybe we should be asking how to not die.

Remember, if you eat right, get plenty of rest, exercise regularly, and don't smoke or use drugs, you will eventually grow old and die.

22 - HOW TECHNOLOGY MADE MY LIFE BETTER

The secret to life is file management. Achieving this is more complicated than you might think. See, no matter what you do, you accumulate things. Things like receipts, web searches, passwords, phone numbers, cool quotes, pictures, things-you're-going-to-do-someday, shopping lists, keepsakes, and little treasures constantly pile up. After a while, you can't find any of it.

So, I decided I needed a system and this is where technology saved me. I began by setting aside a place on the computer to keep all my important stuff. This was more complicated than you might think. When I first started out, I quickly learned that just saving something without specifying exactly where it should be saved, sent that item directly to "someplace". It's just that I will never again know where that place is, forever.

So, I made a file to save things in. This was more complicated than you might think. Right away I ran into the problem of naming a file. Naming things requires significant mental activity. There are several purposes for naming things. One is to identify the named object. In so doing you also differentiate which one of many similar things you are thinking of.

For example, I frequently explained to our children that they all had different names because we assumed they would all be different. My wife thought I was overdoing it. But it turned out that they all were different, and my brilliant insight saved us a lot of trouble later.

Not yet fully understanding the secret to life, I initially named my computer file "Gary". There is really nothing wrong with naming your file by your own name. It's just that, after a while, it got so full there really wasn't much difference between having a file to put things in and not having a file to put things in. I still couldn't find anything. It's kind of like my wife's craft room.

The solution was obvious. You merely create files inside your file. This is more complicated than you might think. Of course, you must name each file, so you can remember where you put the files inside your file. At this point, I can tell you that it is a bad idea to name the files in your file names like "Gary 1" and "Gary 2". It is better to name them something reflective of the stuff you put in the new files.

For example, you could name one of the files inside your file "writing". This is where you would put all the stuff you write. If you never write anything, you would not need this file. If you write a lot, you will discover that you have written so much you can't find the things you remember writing.

The solution is to create files inside your "writing" file. This is more complicated than you might think. You might have writing files labeled "letters", "stories", "articles", "columns", and "books". Later, you might want to add files to your column files called things like "in progress", "current", and "Past Files". Past files might have files inside it labeled by year: "2013", "2014" etc.

Of course, "writing" is just an exemplary category. You may need files like business, home, church, pictures, personal, hobbies, club business, games, collections, books, music, or taxes. Each person will have their own unique set of files, files within files, and files within files that are within files.

This has worked out beautifully for me. I now have so many files that I can't find the file in which I saved the file that contains the file I want. So this weekend I plan on making a file that contains a list of all the files I have, with the files found in each file. Thank goodness for technology, so I can keep it all straight.

P.S. Writing this chapter has been more complicated than you might think!

23 - METABOLISM

"Are you suffering from SCRS? Are your chemical reactions too slow this summer? Is this interfering with your ridiculously-frantic, fun-filled, inappropriate and irresponsible senior years? Or are you still working at the age of seventy-nine in order to afford life insurance and you need better chemical reactions to keep it up? Contact your doctor and ask him about 'Somersukers' for Summer Chemical Reaction Syndrome. (SCRS should not be confused with senior caudal regression syndrome or stupid colorectal surgery. Some patients may die or worse.")

The chemical reactions that happen in your body are collectively called your metabolism. Heat is supposed to speed up chemical reactions, so why am I so lethargic when it's hot? I am proposing a new medical condition called SCRS for summer chemical reaction syndrome.

Actually, there are two kinds of metabolism. One destroys large molecules by taking them apart into smaller molecules. These reactions are called catabolic reactions. (I always remember this by thinking of the word catastrophe. Catastrophes generally take things apart, and so do catabolic reactions. On the other hand, I can't really think of any reason you would want to remember this.)

The other kind of metabolism is called anabolism. Anabolism happens when larger molecules are assembled from smaller molecules. (I wish I could tell you an easy way to remember anabolism, except I usually can't remember the word anabolism myself. I have to look anabolism up. Don't worry about it though. Now that you know you don't need to know anabolism, you probably won't be able to forget anabolism!)

How in the heck does one get two invisible atoms to join to make a larger, invisible molecule? It's hard enough to assemble things with Legos. Well, it turns out to be not much different than getting two Legos to stick together. If you want two Legos to stick together, they must be aligned properly. Then enough force must be applied to make them stick. It's the same with atoms.

Each atom has a kind of shape to it. (Admittedly, it is an electromagnetic shape. It's a little hard for me to mentally envision electromagnetic shapes, so generally I don't try. I just pretend that atoms have regular, geometric shapes like Legos, only more bizarre.)

So, to get two atoms to stick together, we simply must align them properly and supply enough force to make them stick. Exactly how to manipulate and push on invisible particles is the challenge. (Actually, believing in invisible particles is another challenge.)

What we could do is put all the atoms in a closed container and shake them up real hard. The harder we shake, the more often they hit together. The more often they hit together, the more likely some of them will be aligned just right to fit together. In addition, the harder we shake, the harder they strike together and the more likely they will stick together.
It's a tradeoff.

If I throw something hard enough, I can make it stick to almost anything. (I once stuck a Lego to the wall, but that is a different scientific phenomenon.) The more properly the two are aligned, the less force I will need to make the two fit together. Just like the Lego in the wall, one can overcome poor fit with force. Proper fit can reduce the force needed. We don't really shake things up in a box, of course. Instead we heat the atoms together causing them to move around violently as if shaken. The hotter the temperature, the harder and more frequent the impact, the more likely we are to get an anabolic reaction.

Therefore, you'd think I would have more energy when it's hot. I intend to do some research on this topic as soon as it cools off a little and I get more energy. It's a narrow window though. I get so tired when it turns cold.

SERIOUS IDEAS

You really can't trust most of the stuff in this section. All of these topics were generated in a brain. I have discussed previously how unreliable brains are for thinking. I mean if your thoughts are nothing more than the sum of all the chemical and physical reactions of the universe, leading to the earth, leading to living things, leading to humans, leading to brains, leading to neurons, leading to neurotransmitters, all randomly firing according to predetermined laws over millions of years, who can trust'em?

Still, it might be useful to explain a little about some of the ideas that have developed the ideas that ideas don't exist. If for nothing more than a good laugh.

Most people don't like to think about ideas too much. You really can't think about thinking unless you think about thinking about something. Ideas are sort of an abstraction of an abstraction. Abstractions are anything you can't hold in your hand so they are the opposite of the material world. However, an abstraction of an abstraction might be something physical. Hard to know.

I think it is one of the strange ideas of science that nothing exists other than the material world and then we immediately invent abstract ideas to describe how the material world operates. I deny any knowledge about how people used to think anciently. I am not that old. But according to scientists, people were a bunch of ignoramuses in the old days and believed in things like invisible spirits, souls, essences, vitality, force, and substance.

Of course, that is all behind us now and we believe in invisible things like forces, fields, momentum, spin, potential, pressure, tension, energy and velocity. Oh yeah, and math. We believe in math rather than spells and righteousness. I mean, if you can't hold it in your hand, is it material?
Most of these modern ideas have a history founded in many of the ancient ideas.

However, people tend to be even less interested in history than they are ideas and the history of ideas has got to be some kind of nadir, below which it is impossible to go lower on an interest scale. So, if you want you could skip chapter twenty-four to thirty which will discuss nothing material whatsoever. Just a bunch of machine generated thoughts.

24 - THEORY/THEOREM

I have a new theory. I should point out that it is not a theorem. You may wonder what the difference between a theory and a theorem is. You probably don't wonder but, just in case you do I am going to tell you.

A theory is a generalized explanation for how the world works. It is often abstract, generalized and contemplative, but also cool. I have found that forming theories can cover for a lot of staring off into space and doing nothing!

It's probably not as good as fishing that way because you have to explain that you are contemplating. If you are holding a fishing pole, you don't have to explain that you are fishing. Still, when someone asks, "What are you doing?" you can just say "Contemplating." as if you know what that means.

Anyway, a theorem involves a lot more work. A theorem is a statement that has been proven based on previously established statements. Right there it gets more complicated than a theory because you must actually know what you are talking about. You have to be consistent with previous statements, or other theorems. And a theorem must be true. I think I'd rather fish.

Another cool thing about theorizing is that you don't REALLY have to prove anything. In fact, theories cannot generally be "proven". All a theory must do is suggest a hypothesis that can be tested. If an experiment is performed and the result does not support the theory, the theory is considered incorrect. In other words, the key attribute of a theory is that it can be shown to be false.

Non-scientists sometimes use the words "theory" and "hypothesis" interchangeably. But you don't really have to test a hypothesis to have a theory. By carefully avoiding hypotheses, one can talk theoretically with little actual effort. If no one ever does the experiment, they can never be proven wrong. This is one of the major advantages of being a political scientist.

Hypotheses can be useful in their own way though. Probably the easiest way of coming up with a "hypothesis", if

you are sure you want to go to that much work, is to write your idea out as an "If,. . . then" statement. For example, "If there is artificial intelligence then there must be artificial stupidity." Or, "If you torture data correctly, then it will confess."
So, my new theory is this. "In theory, there is no difference between theory and practice, but in practice there is". This is based on a lifetime of observations that in theory everything is clear but nothing works. However, in practice everything works, but nothing is clear. Now keep in mind that this is a theory, not a theorem, and I don't need to prove this at all.

However, my new theory does suggest a testable hypothesis. For example, an "If . . . then" statement might look something like this. "If everything is clear but nothing is working, then the idea is a theory." This might be useful in evaluating government programs. It could also be used to classify mechanical systems. "If everything is working, but nothing is clear, it is practical." Another way of phrasing this latter idea as a hypothesis is "If it ain't broke, then don't fix it." If everything works and everything is clear, then it is a theorem.

I thought about trying some experiments on this theory. As I tried to apply practice to theory, I discovered that nothing works and nothing is clear. That always seems to be the problem with my theories! Or maybe that's the problem with my practice. Now I'm confused. Anyway, all this is speaking theoretically, I suppose. Or am I speaking hypothetically? I always get confused about that. Of course, that's Roget's fault because he lists them as synonyms for each other. One thing is clear. I am not speaking theoremetically.

25 - WRITING

I have been reading in the journal "Advances in Psychiatric Treatment". I was just looking something up for a friend. In doing so I stumbled across an article on "The emotional and physical health benefits of expressive writing." My first thought was what is "unexpressive" writing? The best I can make out is that "expressive writing" must last for several minutes and must be about a painful or stressful event in one's life. That makes this book "unexpressive". The reading is probably "expressive."

It's not that writing for several minutes at a time isn't a painful and stressful event for me. See, my left thumb hurts. It especially hurts when I play the guitar, mandolin, or banjo. Someone tried to tell me I have a repetitive motion injury, but that can't be right. That thumb hurts whenever I do any kind of work at all, and that is hardly ever. It is especially painful for me to write with my left hand, but that is probably because I am right handed.

Anyway, since I write something almost every day, expressive or not, I was intrigued. True, technically I am not writing. I am typing. I was curious if keyboarding would have the same effect and they never said. I also wanted to know if writing would do anything for my gall bladder. I mean, exercise is good for the heart, brain, lungs, and blood pressure. I got all that. What can I do to protect my other organs like my liver, pancreas and gall bladder?

According to Karen A. Baikie and Kay Wilhelm of the Black Dog Institute and School of Psychiatry at the University of New South Wales, writing about stressful events for several minutes, after they occur, can improve both physical and mental health. Apparently writing about painful and stressful events before they happen doesn't help that much. In fact, it may even trigger a painful or stressful event.

I am not sure if studies on black dogs are directly transferable to humans or not. However, you must admit teaching black dogs to write at all is a pretty impressive accomplishment! Many of their entries demonstrated a surprising depth and range of emotions.

The dogs were encouraged to, "Really let go and explore their deepest emotions and thoughts, and to tie their topic to their relationships with others, including parents, lovers, friends or relatives; to your past, your present or your future; or to who you have been, who you would like to be or who you are now."

Wow! This is just like book except the black dogs had a far broader range of expression and emotions. They often reflected on just how stressful and painful their lives had been. I just can't seem to fake that kind of thing. Their further instructions were not to "worry about spelling, grammar or sentence structure." My wife, who is my first proof-reader will tell you that I am wildly successful in this category.

Anyway, in the short term, participants found that such writing depressed the heck out of them. I sort of feel this way each day as I finish writing. However, the long-term effects seemed to be improved liver function and several other physical benefits from blood pressure and respiration to the immune system. They also showed improvements in various and sundry, social, psychiatric behaviors.

It was the liver function that really caught my eye, however, as the gall bladder has some relation to liver function. If this is truly the case, then "expressive" writing activities hold out some promise for prophylactic gall bladder health, at least in black dogs.

I found this discovery tremendously exciting until my wife reminded me that I had my gall bladder removed several years ago. Still, I feel much better about writing these science books now. Apparently, it is good for my health, if not for yours. If you are now feeling depressed, try writing an expressive letter to the editor.

26 - NEUTRALITY

Scientists talk about things differently than normal people. I don't think it's intentional. Maybe the same sort of thing happens in other disciplines. After a while, language simply starts to mean different things to different people based upon their shared experiences.

The reason we can communicate at all is because we share some similar experiences. We think the sky is blue because we were told it is blue. If someone had told you as a toddler that the sky was red, like my sister told me, growing up normal like I have would be a miracle. When I say that a destination is a block away, most of us have some shared experience with what a block means.

However, when we make the decision to delve deeply into any subject, the experiences we have compared to the experience of others can be different. At that point language can start to have different meaning from that of normal people. (I am assuming here that only abnormal people delve deeply into subjects.) A good example of this is watching ESPN. What are those guys talking about?

This column is a good example of the strange use of language. I maintain that I am normal despite harboring earlier beliefs that the sky is red. I then say that only abnormal people delve deeply into subjects, which I have sometimes done. My saving grace, as far as being normal is concerned, (assuming I have any saving graces) is that I think I might have ADD and seldom delve deeply into anything before I get distracted.

Now, what was I saying? Oh yes, scientific language. It goes beyond just the use of weird Latin terms like "Macrocanthorhynchus hirudinaceous", or chemical jargon like "six hydroscopic ambaphascient tetra halide". Sometimes understood concepts in science become obscured in everyday usage.

For example, most people know that when they speak about "neutrality" they are talking about something in the middle between two extremes. In fact, that may not be what "neutral" means at all for scientists. In chemistry, a neutral solution is a solution that is neither acid nor base. Neutrality is sometimes achieved by mixing acids and bases in such a manner that they cancel each other out. However, it can also be achieved in a solution that simply does not ionize. In other words, a neutral solution is not in a state of being acid or base. It's not stuck in the middle someplace.

In physics, neutral refers to the lack of any electrical charge. Neutral is not half positive and half negative. It is non-electric, at least now. In the physical world, a car in neutral is not half way between fast and slow, or forward and back. A car in neutral isn't in any gear at all.
I guess most of the time these little differences don't matter. If I don't know that the "five hole" is the gap between the goalies' knees in hockey it doesn't make much difference. I don't watch hockey. But sometimes the nuances of language can be worth examining.

When we use the word neutral to mean something or someone in the middle, it may be very misleading. What did it mean to not take sides in World War II? Switzerland was neutral. Did that mean they were in the middle between good and evil, that they aren't in either condition, or that they don't know the difference? Since good and evil depend upon a standard, like gears or electrical currents, does that mean Switzerland had no standards of good and evil?

What does it mean when people maintain that government must be neutral when it comes to standards of behavior such as are found in religion, ethics, or morality? Does that mean government must take no stand, or does that mean there are no standards for government? What if the sky really is red and we've all been told wrong?

27 - ANXIETY

I always hesitate to write columns about psychology. I might be opening a can of worms that could take a very bad turn towards self-incrimination. Normally, being a parasitologist, a can of worms is right up my alley. However, psychology is very different than parasitology in that things are never what they seem.

I mean a worm is a worm. But even paranoid columnists have enemies. That is why, now that I have made it clear that I am psychologically normal, I am telling you right up front that I refuse to answer any questions about this column on the grounds that it might incriminate me.

I have been so busy trying to save the world the last year or two, with absolutely no success, that I have fallen way behind on goofing off. My wife and I have different views on the value of goofing off, so I am always on the lookout for scientific data to support my position. My position, of course, is that it's too late to save the world and I can do a better job of not saving it if I am rested.

So, I've been catching up on my summer reading. Normally I don't read science stuff in the summertime because I am sick of it after a school year of teaching and doing research. Then it dawned on me that I also don't read science in the winter because I am too busy doing research and teaching. So, right after I finished the Calvin and Hobbs book, I turned to a stack of papers on my desk

Taylor's Manifest Anxiety Scale (No relation to Stuart Taylor. He is just an anxiety carrier.) is a test that has been used since the early 1950's to measure how much anxiety a person is experiencing. It's simple actually. A person is asked to consider thoughts about themselves such as, "I often worry that something bad will happen." If you say, "that's me", it adds to your anxiety score. If you say, "not me", it subtracts from it.

My problem with these tests is that I never like the alternatives given. I would rather have a range of more honest answers like, "It always has!" or, "Only on days that end in Y!" However, Taylor has apparently been giving these tests to kids for over fifty years. The paper I read compared the kids of 1948 to kids in 1989. Here are some results.

I wake up fresh and rested.	1948 - 74.6% 1989 - 31.3%
I'm under a lot of tension.	1948 - 16.2% 1989 - 41.6%
Life is a strain for me.	1948 - 9.5% 1989 - 35%
I worry a lot	1948 - 22.6% 1989 - 55.2%
I am afraid of losing my mind.	1948 - 4.1% 1989 23.4%

    Now it's obvious that for the last forty years our children have been growing more anxious. Other data suggests this trend has continued into the present. For example, another psychological test (the Rotter Internal-External Locus of Control Scale) has shown an 80% rise in external control for children and young people into their twenties from the 1950's to 2002.

Studies indicate that depression and anxiety increase in people when they feel as if they do not have personal control. In today's world children under 65 have very little personal control of anything.

    As significant as this may appear, and as concerning as this is for our nations young people, I found it even more disturbing that I do not awake fresh and rested, work under a great deal of tension, experience significant strain much of the time, have more than my fair share of things to worry about, and have been accused of losing my mind. But that is just my enemies talking.

    Forget the kids! I am suffering from anxiety. So, psychologically, I conclude that anxiety and depression are the new normal. Hence, I am normal. No more questions!

28 - AGNOTOLOGY

Humans engage in some fascinating activities. Most of them make no sense, but I guess that is what makes them fascinating. Like the health industry. What are we trying to do? What is health but the absence of disease? What is disease? Why, the absence of health, of course! How will we know when we get there?

The education industry is as poorly defined as the medical industry. What is knowledge but the absence of ignorance? What is ignorance? It is the absence of knowledge. So, we go around and around but never seem to get anywhere. I have a fascinating book on my shelf entitled "Three Thousand years of Educational Wisdom: selections from great documents." It appears humans have been trying to learn how to educate people for a very long time. How do we know when we get there? Oh wait! I know! Testing.

Apparently, we haven't been all that successful since we still widely disagree about both medical and educational practices. Maybe it is time to change our approach. Maybe instead of studying how to put knowledge in we need to study how to get ignorance out. I was delighted to learn recently that there is an entire field of the study of ignorance called Agnotology. Seriously! Wouldn't it be cool to have a PhD in ignorance? I wonder what the final exam would be like.

I am not making this up. Dr. Robert N. Proctor did. That shouldn't disturb you since almost all of science was just made up at one time or another. It's only later when we decide that something is real. The jury is still out on whether Agnotology is real or just something Dr. Proctor made up. Personally, I think the evidence supporting the reality of ignorance is overwhelming. My opinion is based on a small sample size of everyone I have ever known so Dr. Proctor and I could be wrong.

When founding a new field of study, scientists usually begin by classifying categories of the objects of study. Dr. Proctor has delineated three basic kinds of ignorance. These classifications may seem a bit arbitrary, but they are a good beginning.

First there is "ignorance as a resource". This kind of ignorance is typical of a child who is simply lacking knowledge. I tend to think of it as a resource because it is the kind of ignorance that keeps public school teachers and University professors employed. Someone has got to shovel knowledge into those empty heads.

"Ignorance by choice" has both benign and malignant causes. I tend to remain relatively ignorant about sports. Full disclosure requires me to tell you my sons think I am pathologically ignorant about sports. That is an ignorance by choice. I simply cannot keep track of baseball stats and guitar chords at the same time. There is simply too much information in the world for one to know about everything. Even choosing a career limits what one will learn, although most scientists think they know it all anyway.

"Ignorance as a product" is probably the most interesting category. In this case, ignorance is creatively manufactured by someone casting doubt about what you know. There are people who may not want you to know certain things. One ploy would be to convince you that what you think you know isn't true. This inevitably leaves you confused.

This manufacturing of ignorance is common in advertising, politics and difficult relationships. False information is a form of manufactured ignorance. As a benign example, my wife used to ask me how to spell something. When I told her my best guess, she would ask, "Are you sure?" I never was.

I propose that there are many more sub-categories of ignorance such as secrecy, stupidity, apathy, censorship, statistics, and forgetfulness. Is ignorance the absence of knowledge or, as in my case, just a state of confusion? Humans engage in some fascinating activities. Most of them make no sense.

29 - ODD OR EVEN

Have you ever wondered why we call numbers that are divisible by two "even" and all the others "odd"? What makes things like one, three, five, and seven any more odd than two, four, six, or eight that we appreciate? There are just as many odd numbers as there are even numbers, so it's not like odd numbers are some kind of minority.

I think it is just as odd to be even, but I suppose that may be because I am odd. The natural world appears more odd than even. I mean, perfect symmetry is a rare thing in nature. Even when there is a pattern, it is often a little uneven. Doesn't that make it odd? In a world where it is improper to call odd things odd, why are we still using hurtful terminology to talk about numbers?

This is just another sign of the coarsening of American society. Even math has become polarized with hurtful and ugly things being said on both sides. In fact, the odd and even thing is on a par with terms like "he" and "she" or "male" and "female". It's all sexist. I didn't ask to be born with an odd chromosome. It's a birth defect, and polite people shouldn't make fun.

There might be a better system. The Greeks had a promising idea. They didn't have calculators so used marbles instead. Yes, the Greeks claim to have invented marbles because a taw and three aggie ducks were found in the ruins of Pompeii. If you don't have all your marbles, though, you can use little pebbles to follow this along.

Starting with one pebble and adding two pebbles gets you three. Three pebbles can be stacked into a triangle. Since you added two to the first pebble, you now add the three, the next number in sequence, to the three pebbles and you will get six pebbles. Six pebbles also make a triangle. The next number in sequence is four and if you add four to six you get ten which can also be stacked as a triangle. Continuing this pattern of adding the next number of pebbles in sequence will always give you a number of pebbles that can be stacked into a triangle. Therefore 1, 3, 6, 10, 15, 21, etc. would all be "triangular numbers".

Now, what if you start with one pebble, and instead of adding two pebbles, you start by adding three pebbles? That would yield four pebbles, and four cannot be arranged into a triangle. Four must be arranged in a square. From here, add three pebbles again, plus two more for a total of five, and you get nine which is also arranged as a square. If you add five pebbles plus two (seven) to nine you get sixteen.
If you continue this pattern of adding the previous number of pebbles, plus two, you will always get a number of pebbles that can be arranged into a square. Hence these are called "square numbers", but are not to be confused with squared numbers. The road goes on forever, and the party never ends.

An interesting feature of this arrangement is that if you add two triangle numbers together, you always get a square number. And, of course, all square numbers can be divided into at least two triangles.

Looking at numbers this way completely eliminates the concept of odd and even numbers and replaces it with triangle and square numbers. Since some triangle numbers are even and some square numbers are odd, there will no longer be a stigma attached to either. This type of arithmetic is called "figurative arithmetic" which you probably thought was something else entirely, you sexist.

Anyway, adopting this old Greek method would go a long way towards making mathematics more politically correct. It's something that has long been needed, but math is rather rigid and old fashioned.

30 - GIVE YOURSELF A HAND

Most people think the brain tells the hand what to do. It seems to me that the brain doesn't know much about the hand until the hand tells it. How can a brain know what is hot or cold until the hand touches hot and cold things? How can a brain know left and right until the hand learns to put the right shoe on the right foot?

Okay, technically I suppose that is your foot telling your brain what is right and left. Still, the concept is the same. Your body informs your brain about the physical world long before your brain tells your hand what to do.

I bet the first thing you do when you wake up in the morning is hit the snooze button with your hand. That is probably the one time during the day when we have a lot of time on our hands. Now, you may be one of those rare, disciplined, disgusting individuals who actually gets up when you are supposed to. Then the first thing you do is turn the alarm clock off, so it won't disturb the other more-normal people in the house.

The fact is, you probably used your hands for dozens of things before you even uttered a word. Okay, maybe you said a word or two when the alarm went off. But that is a learned response and must be carefully taught. Speaking before thinking is actually a very dangerous thing to do and should only be attempted by specially trained windbags called college professors. Do not try this at home.

Babies learn to reach for things long before they learn to talk. Most of them are successful at picking things up and putting them in their mouths before uttering their first words. Could that be the origin of the phrase a "hand to mouth existence"?

What does all this have to do with science? The history of human invention, which we like to call progress, is mostly the understanding that we develop of the physical world through things we build with our hands. Like the old saying goes, "The things we make, make us".

So, what do we make? Music? Gadgets? Gardens? Golf scores? Apps? Does it make any

difference what we make?   What happens when we become a service society and don't make anything anymore?
Scientific discoveries always involve the making of something; from making the first microscopes or electrical circuits to trying to make gold from straw.   The brain seldom knows much until the body becomes involved.

Unfortunately, modern educational practice has either forgotten this, or has chosen to ignore it.   There is still some drawing, cutting, and pasting in the first few grades, highly valued skills I suppose.   But by and large hand education has disappeared. Today, many even want to do away with cursive writing, although it has been shown to increase reading skill and clarity of thought.

The students are left empty handed.   Making things is time consuming and a material driven endeavor.   More expense is involved than merely reading and writing.
However, reading and writing have limited application, unless used to direct someone's hand somewhere.

In my experience, making things almost always involves making mistakes.   Making mistakes, however, is sometimes the most useful thing one can make because we learn so much when we make them.

I had a professor once who said that if an experiment came out the way we expected, it was boring and was an indication that we already understood the system.   Only when the results don't meet our expectations does it get really interesting!

On the other hand, . . . you just have different fingers. There is no "other hand".   The brain cannot know anything until the hand teaches it.   The brain cannot know that it's "reach can exceed its grasp", unless the hand has had to reach and grasp.

SERIOUS SPIRITUALITY

Science is based on the idea of materialism. Materialism is a philosophical belief that matter is the fundamental substance in nature and all thing can be explained by the interactions between matter.

We've come a long way, baby! Back in the old days, whenever those were, people believed that matter had invisible things like spirits, auras, and personalities. Nowadays we are more enlightened and think we know that matter only has invisible things like energy, forces and fields.

The new invisible characteristics haven't been a bad thing for humanity because they have allowed us to think about invisible attributes with more precision and therefore to quantify some of their effects. But they are still invisible.

One of the problems with materialism (the scientific concept, not the philosophy of consumption) is that the philosophy encompasses many ideas that seem truly significant to humans, yet seem immaterial. For example, scientific materialism would have us believe that even our minds, free-will, and philosophy itself is nothing but the result of the material workings of our minds.

It also reduces many seemingly urgent ideas to meaninglessness, such as freedom, love, democracy, and in fact the entire litany of abstract ideas such as energy, force, and field. It turns out that many spiritual ideas have surprising force. They can cause wars, and they can cause peace. In this world, both are powerful.

Anyway, you may think it somewhat strange to include a section on "spiritual reality" in a book on reality. However, you may be surprised at how much science and spirituality share some commonalities.

31 - THINGS AND MIND

At some point, "things" become "alive". No one knows, for sure, what life is exactly but, only in extreme circumstances, do we have trouble telling something that is alive from something that is not. Exactly when does a stack of material acquire that condition we call "alive"?

Most of us agree that electrons are not alive. Atoms do not appear to be alive. The interaction between electrons and atoms can make molecules like proteins. But most people don't think proteins are alive even though proteins make up cells which humans do think are alive. Then, some people don't think a fetus is alive; yet it is made of functioning, living cells. We don't always agree, but at some point, inanimate things start to be what we call "alive".

At some point, "things" also become a "mind". I don't know of anyone who thinks that an atom has a mind. Molecules do not seem to fit the concept of something with a mind. It is less clear whether a cell has a mind.

This confusion arises because in many ways cells can do things you and I can do with our minds. Many cells move about in a purposeful way, detecting environmental conditions and responding to them. They seem to know what they like, and don't like, and how to meet their own needs.

A series of "things" can become more than just a collection of things if they are connected in a specific way. For example, the steering wheel on your car is connected in such a way that it can alter the direction in which you are moving. The steering wheel does not know what "direction" is, yet it can change your direction.

To do this, it must be connected to something else that it tells what to do. The steering wheel certainly doesn't know how it does what it does. It simply turns a shaft, that turns a gear, that pulls a rod, that shifts the axle. A collection of these "things" has become something besides what any of the things are separately.

Life and thoughts are kind of ghostly things. Life only occurs as lumps of stuff, and thoughts apparently only occur in brains - a lump of stuff within a lump of stuff. We express thoughts in words, but they are often just thoughts. When my wife asks me, "What are you thinking?" it is sometimes hard to tell her. A "penney for one of my thoughts" is probably never much of a bargain.

In the past, humans have treated "thinking", and being "alive", like a mysterious box. You can do something to the box, and then watch the result of what the box does in response. There is still no way of knowing what is going on in the box. If you stick a knife into a living person they may cease to live but you may not have any more understanding about why.

We know a lot more about the pieces inside the box now than we have in the past. But we still don't know what the critical point is when the box ceases to be alive, or to think. Sometimes we can even lose the ability to think while still being alive. I've been accused of that.

In the last century, Kurt Godel, Alan Turing, and others have tried building mechanical minds. This line of exploration has given us computers. Computers are less ghostly than minds because we understand the things with which they are made and how they fit together. But the things they are made of are organized in a very different way than our bodies and brains appear to be.

So, here I sit thinking, and I haven't a clue as to how I am doing it. I don't know if humans will ever be able to understand how things become "alive", or how things can make a "mind" that thinks. We still don't know exactly how, at some point, living things stop living and become just "things" again. When a mind stops thinking, what becomes of the thought?

32 - SOMETHING FISHY

I had to teach a lab on fish recently. I'm not big on vertebrates as I think they are highly over-rated. However, I guess it's necessary for humans to know a little about fish since we happen to have vertebrae too. Don't be embarrassed by that though. Vertebrates only make up a small percent of all the animals. About all you and I can claim is to be interesting, fringe creatures. We add variety to the world! Still, a whole lab on fish seems excessive.

Anyway, the lab led to a discussion of fish bladders. Fish don't really have bladders like you and me. I mean they live in water and can just go in the pool whenever they want. No need to hold onto anything. What they have is an oblong bladder filled with gas. This is located internally just below the spine. By regulating the amount of gas in the bladder fish regulate their buoyancy and depth in the water. Me? I depend on the buoyancy of fat to keep me afloat.

This fish bladder has been known about for centuries. Anciently it was called the vesica pisces: fish bladder. Before scientists there were philosophers. Philosophers are people who sit around and ask questions. The early scientists did this also. More recently, however, scientists seem to have evolved into people who sit around, have all the answers, and must not be questioned.

One question early philosophers asked was, "How do you make an accurate equilateral triangle?" About the only tools they had were a straight edge and a compass. You know, the kind of compass that allows you to draw an accurate circle by placing the point in one position and then rotating the pencil around the point with a set radius.

I don't think they used pencils then because they hadn't been invented yet. But they did discover that if you drew one circle, and then placed the point of the compass on the edge of the circle and drew another of the same radius, you created two overlapping circles. The overlapping portion of the circles made a pointed oblong shape like a football, although the NFL hadn't been invented yet either. They called this the vesica pisces because it looked to them somewhat like a fish bladder.

This alone probably doesn't mean much to enlightened modern folks except that they also discovered that if you draw a straight line directly through the center of the circles the line also bisects the sides of the vesica pisces in half. Then if you draw straight lines from the mid points of the vesica pisces sides to the pointed end of the football you create a perfect, equilateral triangle. If you draw a line from the midpoints of the sides to the other end of the football, you make a second equilateral triangle, upside down.

This explains the Pisces constellation of stars. I always thought it looks more like a V than a fish. However, come to think of it, a fish is sort of shaped like two equilateral triangles attached base to base. In fact, a vesica pisces, with the sides extended slightly on one end is the universal symbol for a fish, and the early Christian church.

I must admit that the triangle, as well as the number three, are significant concepts, in spite of the fact that I hated trigonometry. It takes three legs to make a stable chair, three plaits to make a braid, three tenses to express time, and three primary colors. Is it an accident that the vesica pisces is a symbol for the Holy Trinity?

Who knew that geometry and trigonometry grew out of fish bladders? Much of our modern world is based on the number three and the triangle. I mean there are three little pigs, three blind mice, three little kittens, three bears, three wishes, and three strikes and I'm out.

33 - VALENTINES

I was a soldier from 1966 until 1968. I left the military with mixed emotions: joy and gladness. Don't misunderstand me. I am proud that I served my country, and the experience was very beneficial personally. It's just that I am not really very strong, courageous, or obedient. I think, by the end of two years, both the U.S. Army and I had come to grips with that. We parted ways on pleasant terms.

I mention this only because it illustrates an important point often missed by scientists. Since science is the study of our material world, scientists tend to be pathologically focused on the material part of physical objects. Now, it is the nature of physical objects, so science tells us, that no two objects can occupy the same space at the same time.

Therefore, the study of matter is the study of diversity. When cataloging one element, one force, or one animal from another, we are breaking the world up into smaller, separate segments. Categorization has been useful in many ways, yet harmful in others.

By contrast, many important ideas and concepts are abstractions. Abstractions are not physical objects that can be held in your hands. So abstractions can exist in the same space at the same time. Some people say abstractions aren't real or important. Yet love, democracy, and beauty are very important, but are abstractions.

For example, I experienced both joy and gladness upon discharge from the military. You probably have the same mix of feelings every Friday in the late afternoon. People can be curious and appalled simultaneously. I have even felt free and guilty simultaneously!

Many abstractions accompany each other at the same time and in the same place. Young lovers are often excited and confused simultaneously. I am assured that such emotions as love and irritation can occur simultaneously in wives. I have heard this from a number of sources, although some are more trustworthy than others.

Another important limitation to the study of the material world is that the same material object cannot exist in two different places at the same time. However, the same material object can occupy the same space at two different times. In fact, this latter idea is obligatory.

Luckily for reality, abstractions do not have these same limitations. In fact, the situation is almost reversed. Two abstractions can exist at the same time in different spaces. How else could one explain romance as being two different bodies feeling attraction and affection towards each other simultaneously. What about beauty and sadness?

Unlike the material world, it is not necessary for two abstractions to occupy the same space at different times. Abstract ideas can move about from time to time on their own. Abstractions such as "emotions" don't just sit there in the same place all the time. Love can be lost or, at least, temporarily misplaced.

Anyway, I think all the attention being paid to diversity in our world is the end result of the scientific revolution where everything has had to be categorized into separate physical spaces. So, humans have imbued abstract ideas like race, culture, political parties, and religions with the material characteristic of separateness.

Doing this is detrimental to achieving unity of ideals and goals because unity, an abstraction itself, can apparently only happen in the strange, spiritual world of abstractions. Only in the non-material world of abstractions can ideas such as shared feelings and attitudes unite at the same time and space, or occur at the same time in two different places. Perhaps this points to one of the roles that religion can play in our modern world. If we don't see religion as a material that separates, perhaps we could see it as a unifying force.

Abstractions such as love, unity, commitment, and faith, are things that unite man and wife and parent and child, as well as communities and countries. More love!

34 - EASTER

It is fascinating to me that one of the least scientific events of recorded history has been the most significant event in the history of science. The resurrection of Jesus Christ is without precedence in human history. Scientifically it is impossible and miraculous, a word scientists eschew. Yet it is Christianity which gave birth to modern scientific method and ideas.

Christianity did not invent reason, knowledge, or the idea that the universe is a logical, rational place. Pythagoras, Thales and many other early philosophers postulated a universe that was understandable according to natural law. Plato taught there was a natural law. Algebra was invented through ancient Islamic culture. However, these early ideas were ignored for centuries during the dark ages. But they were recorded, kept alive, and eventually rediscovered by early Christian scribes and thinkers in Monasteries.

It was in the early Monasteries where methodical experimentation with methods of manufacture, agriculture, and other practical matters were first attempted. The monasteries gave birth to the first Universities, which were all religion based institutions. Science was born of these efforts.

Christianity resurrected the idea of an orderly universe because it believes that such a world has a rational cause: God. The Judeo-Christian ethic teaches that the world operates according to divine reason. Since we are created in God's image, and God is a logical, orderly, reasonable God, we must also possess these attributes. Of course, like with science, this is also an assumption. Reason always begins with initial assumptions.

Until recently, when other cultures have adopted western scientific methods, the overwhelming majority of scientific discoveries have been made by people who had absorbed the Judeo-Christian ethic. This ethic is the idea that if man studies the way the world works hard enough, mankind will better understand the mind of God. Ninety percent of all scientific discoveries of the last several hundred years have been made in countries heavily influenced by the

Judeo-Christian ethic. The list of scientists who were devout Christians is long and fundamental to our modern understanding of the world.

The atheist scientist is a very recent phenomenon, perhaps a couple of hundred years old at best. They have gained prominence in modern culture due to the militaristic proselytizing of a group of men who use extremely unscientific methods and arguments to attack religion. A careful study of their words and methods reveal that their arguments are very little about science or religion and more about a way of life they wish to pursue.

The universal laws discovered by science, and our ability to predict and control using them, have been a great gift to all of mankind. It isn't necessary that you believe in God to have the advantage of gravity. Electricity works just fine for atheists.

Much of modern medicine was developed by overtly-religious individuals of the Judeo-Christian heritage, but is a benefit to people of all faiths. If humans will avail themselves to education and use knowledge in a beneficial way, science can improve the lives of all mankind, in spite of man's varying religious beliefs.

The unscientific resurrection of Jesus Christ has blessed mankind in ways little understood by modern man. The influence of the Son of God goes unrecognized by much of the world, just as his religious role is unknown and unappreciated by huge portions of the world's population. For those who already recognize his role as Savior and redeemer, it might be interesting to understand his unique role in our way of life through scientific discovery.

35 - A UNIVERSAL HOLIDAY

Lewis Carroll imagined a world in which everything happened differently than it does here. That isn't so surprising. If there are other worlds, wouldn't they be different than ours? I always smile when I read about human attempts to find extra-terrestrial life. What makes us assume we would even recognize extra-terrestrial life if we saw such?

Scientists often disagree about what is alive and what isn't. One often-cited characteristic of life is that living things can reproduce themselves. But so can computer viruses. What if another life form was nothing more than a series of electrical pulses? Would we recognize it as living? If computer viruses are alive, perhaps we should organize a group of people for the "ethical treatment of computer viruses" (PETCV).

Scientists believe life is universally carbon based because we believe the laws of the universe are uniform throughout the universe. We also believe they have been the same throughout time. We even assume these laws will never change. Many of these laws are so precise they appear to be able to be expressed in the language of mathematics. Of course, these assumptions are impossible to prove, except that we haven't found any exceptions yet.

These assumptions are little short of amazing. The presently-accepted theory of the origins of the universe postulates a nearly infinite explosion. Can a violent, chaotic explosion give rise to an orderly universe? To propose that the very explosion was caused by laws, that must have existed prior to the explosion to cause it, and then to not ask what the source of these laws are is a leap of faith of the first order.

Christianity did not invent reason, knowledge, or the idea that the universe is a logical, rational place. Pythagoras, Thales and other early philosophers postulated a universe that was understandable according to natural law. Plato taught there was a natural law. Algebra was invented through ancient Islamic culture. However, these early ideas were ignored for centuries during the dark ages, but were recorded, kept alive, and eventually rediscovered by early Christian thinkers.

The book of genesis tells us, "In the beginning God created Heaven and earth." No other religious book besides the Judeo-Christian Old Testament, addresses the origin of the universe, let alone details the events. Christianity resurrected the idea of an orderly universe because it believes that such a world has a rational cause: God.

The Judeo-Christian ethic teaches that the world operates according to divine reason. Since we are created in God's image, and God is a logical, orderly, reasonable God, we must also possess these attributes. Of course, like with science, this is also an assumption. Reason always begins with initial assumptions.

Until recently, when other cultures have adopted western scientific methods, the overwhelming majority of scientific discoveries have been made by people who had absorbed the Judeo-Christian ethic. This ethic is the idea that if man studies the way the world works hard enough, mankind will better understand the mind of God. Ninety percent of all scientific discoveries of the last several hundred years have been made in countries heavily influenced by the Judeo-Christian ethic.

These universal laws, and our ability to predict and control them, have been a great gift to all of mankind. It isn't necessary that you believe in God to have the advantage of gravity. Electricity works just fine for atheists. Much of modern medicine was developed by overtly-religious individuals of the Judeo-Christian heritage, but is a benefit to people of all faiths. If humans will avail themselves to education and use knowledge in a beneficial way, science can improve the lives of all mankind, in spite of man's varying religious beliefs

However, science would not hold the place it does in our lives without Judeo-Christian scholarship. Even if that faith were to prove erroneous, it has still been the source of knowledge and wisdom. So, Merry Christmas everyone, from Science!

36 - WHAT ARE PEOPLE FOR?

Since my retirement from the University, I no longer have large research grants and extensive laboratory space for my research. Well, actually I've never had any large research grants, and most of my research had to be done in the basement of the old Wubben Hall because all lab space was needed for classwork. Oh, yeah, there is a basement in the old Wubben Hall. You don't want to go there.

Anyway, I now have time to speculate on some less-important scientific questions that have defied analysis. Like, "What are people for?" I am always being asked, "What are mosquitoes for?" or, "What are centipedes for?" I think what people mean is, "What good are mosquitoes and centipedes to people?"

I have read that some people think there are too many people. How can we decide there are too many people unless we know what people are for? If the job people are for is getting done and there are people with nothing to do, then I suppose we should get rid of a few of them. I have a couple of suggestions.

I'm reminded of the time, several years ago, when the government was telling us that there were too many farmers. We had to get rid of some of them, so we could be more efficient. Interestingly, I never heard a farmer express that concern. Too, I never heard a professor of agriculture opine that there were too many professors of agriculture. However, if we had fewer farmers, wouldn't we need fewer professors of agriculture, and a smaller Department of Agriculture?

Where did all those farmers go anyway? They went to cities of course. Apparently, our cities need a lot more people than our farms to do whatever "people are for". I sort of know what people are for on a farm. I am not as sure why we need people in the cities, but there must be some reason because all the farmers went there.

In 2014, it is starting to look like we have too many people in the cities. With all the efficient farms, automated factories, labor-saving devices, robots, shorter workdays, longer vacations and early retirement, it sort of appears that people are just for goofing off. My wife thinks that's what I've been doing for years. . .

Maybe there are too many people in the cities and on the farms, both. If we don't know what they're for, it's hard to tell. Of course, if we think there are too many, I suspect there are further inhumanities on the horizon.

Then again, I may be going about this all wrong. Maybe I should be concentrating on the question, "What good are people to other people?" If we judge other living creatures by this standard it seems only fair to judge people in the same way. Several testable hypotheses come to mind.

If it weren't for the people who invented computers, I would still be writing my columns in long hand. Therefore, my editors would be unable to read my columns, so you would never have to suffer through this stuff! Okay, I guess that could be taken as a pro or a con.

But without people, we wouldn't have electricity, cars, air planes, indoor plumbing, television, cell phones, or Facebook. Nor would we have Shakespeare's plays, Rembrandt's paintings, or Beethoven's symphonies. Ideas and creations come from people. In fact, as far as I know, ideas never come from anything but people.

There is at least one other thing that humans "might be for". Almost all people are fellow humans, friends, neighbors, parents, children, citizens, children of God, and kind people who read my books. So, my scientific hypothesis is, people are for taking care of each other through caring, creating, and having better ideas. That means I just have to figure out what all those people who don't read my column are for.

37 - STRATEGY

Newlyweds shouldn't play Monopoly. I'm not sure why they would want to, I simply maintain they shouldn't. Just take my word for it, okay? Actually, I don't think anyone should play Monopoly who wants to have peaceful relationships. You can think of this as free, post-Valentines Advice.

My problem is that I don't have any sense of strategy. Wait. Maybe it's tactics I don't understand. I get confused, but my family assures me that I don't have any sense. Of strategy, I mean. I haven't won a game of anything since the fall of 1966. I was ahead in a Monopoly game that October when my wife accidentally knocked over the game as she got up from the table.

The problem I have with most games is that people who play them generally want to win. Now, I might enjoy winning, but having never done so, I am not sure how it would feel. Consequently, I value winning less than others seem to, especially my wife. Since winning doesn't seem all that meaningful to me, and it seems tremendously important to others, I just figure why not make them happy. I don't let others win. That wouldn't be any fun. I just always seem to lose.

However, I recently attended an art show featuring some amazing 18th century paintings. One of the paintings was a picture by Carl Bloch of Christ emerging from the tomb. The detail was astounding, and part of that detail were two dice left on the ground from where the guards had been casting lots.

Now scientists know a little about dice because we have to use statistics and probability. This puts us in a unique category of liars. But since Monopoly is played using dice, it also suggests a certain strategy. Well, assuming I know what the word means.

It struck me that the only way to occupy any square in Monopoly is to throw the dice and move that number of squares. Except jail. You can land on jail, or be sent there by landing on the Go-To-Jail square. You can also throw too may doubles, draw a Chance or a Community Chest card, or simply visit.   The point is that jail has got to be the most frequented squares in the entire game because there are so many ways to get there.

That means people spend more time leaving jail than any other single space.   The chance of landing on any property, then, is highest for the properties in the spaces after Jail.   The first five spaces are: St. Charles, the Electric Company, States Ave., Virginia Ave. and the Pennsylvania Railroad.   To land on any of these properties from jail, a player would have to throw two dice that would yield numbers between one and five.

One, of course, is impossible.   There is only one way to roll a two or a three and only two ways to roll a four or a five.   However, there are three ways to roll a six, seven or eight.   Some ways of rolling these numbers involve rolling doubles which makes the possibility for rolling these larger numbers even higher.   In fact, taking doubles into account the most likely moves after leaving jail would involve rolling totals of six, seven, eight, or nine.   A move of either six, eight or nine would land you on St. James, Tennessee, or New York.

The same reasoning applies to the next move after escaping Jail.
The next most likely rolls would also move you in the neighborhood of from six to twelve spaces.   You'd be likely to land on Kentucky, Indiana, or Illinois.   In fact, statistically, the most-frequently-landed-on-property in Moncpoly should be Illinois.

I'm not sure if this insight is spiritual inspiration, strategy, a statistical lie, or useless trivia.   But if you want to win at Monopoly, buy orange.   I'll probably never know if that is really true because since October of 1966, I don't play Monopoly.

38 - GRATITUDE

I'm grateful for Thanksgiving break from school. I'm also grateful for Christmas break, fall break, spring break, and summer break. I know most people don't get those breaks, which makes me even more grateful. I don't bring this up to taunt anyone. That wouldn't be sports-man-like. Having never been much of a sportsman however . . . .

This may sound like I don't love my job. I actually do. Dealing with intriguing ideas every day, trying to discover more about how the world works, working with hungry, young minds . . . well two out of three ain't bad. Seriously, it's a great job! It just cuts into my time for more important things, like sleep. I've tried sleeping on the job, but the students keep waking me up.

Actually, the reason I am so grateful for my job is because I have to work very hard at it. Yeah, I get a lot of breaks but, over the years, I have spent most of them doing research, grading papers, writing articles, or preparing lectures.

The truth is I enjoy working hard. I think work might be what creates my sense of gratitude. It definitely isn't the size of my paycheck. (Just kidding, Tim.) For the most part, I don't like to be given things. I am even a little uncomfortable with Christmas and Birthdays. I believe, for the most part, people are most grateful for the things they have had to work hard or sacrifice for.

In fact, I don't even like compliments. When we were first married and someone would complement me, I would sometimes not say anything at all because I didn't know what to say. Well, at least that's what happened the one time I got a compliment.

My wife taught me how to be civilized though. She explained that I shouldn't deny the compliment, nor ignore it, because it was probably meant sincerely. All one needs to do is simply acknowledge the kindness by saying, "Thank you." If I seem partly civilized in my later life, it is through my wife's efforts.

It's interesting how "insignificant discoveries" take on an

importance when they come at the end of two years of late nights in the lab. Others who have not spent the late nights may fail to see the significance of your modest accomplishment. Like my wife. Or maybe I should appreciate all of her late nights for the family more.

There seems to be a lot of research, in recent years, on the effect of gratitude on human happiness and success. (But I bet people who do research on gratitude don't stay in the lab until two in the morning.) Of course, all of the research cited is apparently from the same two papers. It seems clear from the number of entries on the internet citing those same two papers that gratitude makes a person happier and healthier.

In fact, one researcher goes so far as to say that gratitude is what gives life meaning. Does that seem a little unscientific to you? I mean, who's to say what one's life means? Couldn't a person be grateful but their life still be more or less meaningless? I know I spend a lot of time trying to explain what I actually mean. I don't think dead people get a chance to explain what they meant much.

I think these studies are cases where scientists may have overstated the meaning of their data.
Gratitude is a bit of a conundrum anyway. If gratitude makes one happier and healthier, and if it is hard work that makes one feel grateful, then perhaps its work that makes people happier and healthier, not gratitude at all.

Anyway, I am grateful for the hard work that is a requirement for my job. I am also grateful for Thanksgiving break. After this column, I will be even more grateful if I still have my job next week.

39 - ARROGANCE

Most scientists are humble by nature. Well, at least they are humbled by nature. I admit there are some arrogant scientists. Arrogance seems to be a part of our modern culture. I used to be arrogant. However, that was back when I was nineteen, so it doesn't count. Everyone is arrogant at nineteen.

Without question, arrogance didn't originate with modern science. Philippus Aureolus Theophrastus Bombastus von Hohenheim said, back in the 1500's, "I am Theophrastus, and greater than those to whom you liken me; I am Theophrastus, and in addition I am monarcha medicorum and I can prove to you what you cannot prove... Let me tell you this: every little hair on my neck knows more than you and all your scribes, and my shoe buckles are more learned than your Galen and Avicenna, and my beard has more experience than all your high colleges."

You might note that his fourth name was Bombastus. Besides being arrogant and bombastic, he is mostly known for being wrong. He gave his patients engraved talisman for cures and believed he could cure anything with mercury, sulfur, or salt. Those who didn't die from his treatment always felt better after his cures, so his patients supposed they had worked.

Bombastus did make a couple of contributions, although none so great as to justify his self-esteem. For example, he named the element zinc after the pointed shape of zinc crystals. Zinc means "tooth-like" or "pointed" in German. He noted that, when acids attacked metals, a gas is produced. This is probably the first observation of hydrogen.

With no evidence whatsoever, he came up with the idea that there was a specific medicine to cure every ill. This is still a common belief in today's world. It should be simultaneously noted that he is also credited with the wise old saying with which everyone is familiar, "Dosis facit venenum." "The dose makes the poison." This means that small doses of poison may be harmless and large doses of harmless substances may be lethal.

You might wonder why I am writing about such an odd scientist whose contributions are so marginal. I kind of wondered that myself. I think it's because of the seemingly increasing and unseemly arrogance that runs rampant in our modern world. It seems especially out-of-place in highly educated men and women. Although I suppose arrogance could be an occupational hazard for those who become highly educated, regardless of the field of specialization.

I think this arrogance in our modern world stems from specialization. It's a natural result of the industrial revolution and manufacturing technology where everyone started becoming expert at "just one step in a process." Now we have scientists who know everything about genes but have no idea about the animal in which the gene is located.

In the past, everyone did a little of everything so no one was embarrassed to do anything. Now only the good singers get to sing, and only the good dancers get to dance. Everyone else is embarrassed to even try. It's a shame, really, because I can bust some moves in my own living room.

Besides missing out on a lot of fun by not participating in things they are not special at, specialists sometimes want to protect their specialness. There are lawyers who think that farmers shouldn't get to make laws and free-thinkers who claim that anyone who isn't a free-thinker is a slave-thinker. Politicians don't think the average person can understand politics and the average person doesn't think politicians are average people.

There are even a few scientists who think they are smarter than everyone else. They say so publicly in unreasonable, illogical, and pedantic ways. Some are almost bombastic. Richard Dawkins comes to mind. My wife thinks I'm being arrogant to criticize all the arrogant people. I don't really mean to offend anyone. It's just that when it comes to specialization dosis facit venenum.

40 - HOW THINGS WORK

Things generally only work one way. If you reach into a piece of equipment that is working perfectly fine and randomly grab a wire or a piece of something and rip it out, the most likely outcome will be that the equipment will no longer work.

It is certainly true that you can sometimes keep a piece of equipment functioning when one-part breaks or stops functioning. Take my truck for example. Please! Its basic function is to get me to my bee hives and back. It does that mostly alright.

The inability to unlock the passenger-side door from the outside is just a minor irritation. The fact that once I open the passenger-side door, the cab light won't go off for sometimes up to half an hour is only a problem after dark. So, half the time I am fine.

It's actually good that the emergency brake handle is broken because it keeps other people from driving my truck when the brake is on. They don't know how to get it off. It doesn't matter anymore that the tape player doesn't work because no one makes tapes. The radio is just an irritant anyway. Oh, and the brake pedal is a little slippery since the rubber came off.

I must admit, though, that all of this is just a harbinger of things to come when the truck stops working entirely. I remember when my last truck finally stopped after two motor replacements and an unknown number of miles (the odometer was broken). I asked Mueller if he thought a new motor might be possible. He paused for a few seconds. Then in the tone of a very sad physician he advised me, "Gary, let it go."

Our bodies and cells are sort of the same way. We can get away with taking some things off or out, and some organ systems can stop functioning at peak levels. We know that we might keep going for a while but that things only work in one way. When that way is changed, it means things will eventually quit working.

Like a lot of scientists, I had a miss-spent childhood. Unsupervised a lot of the time, I took a lot of things apart. My room had a lot of radios and clocks, at least one old pump, and assorted plastic toys that were more interesting to disassemble than to play with.   The problem was that I was seldom able to reassemble any of them.   Oh, alright, I don't ever recall successfully reassembling anything.

I put together a robot kit on New Year's Eve a few years ago.   (This probably tells you something about my social life, and my poor wife.)   Anyway, the robot had an infrared sensor and was supposed to avoid obstacles.   Mine attacked obstacles, face planted in them, and then continued spinning gears trying to shove the wall around.

It's generally true that things only work one way.   If, in assembly, you have pieces left over, something is probably not going to work right.   These experiences have sort of left a mark on me.   I'm always a little reticent about taking things apart, and very careful when building new things.   I don't see that the future is going to disprove the past   Like the law of gravity is going to go away?

Anyway, that is why I'm a little worried about the coming year.   Things in our lives have seemed to work pretty well for a couple of hundred years.   Recently, though, we have taken some things out and put some new things in.   My experience has been that this just doesn't always work out too well.

I suppose I could fix my truck and replace all those broken parts.   It might be cheaper than buying a whole new truck, especially one with a bunch of things I really don't need.   Or maybe, like my old truck, I should just "let it go."

41 - ENDINGS

Well, it's getting close to the end of the year. It's comforting to know when the end of the year comes. It's always after 365 days. You can prepare for something like that. It's predictable.

I could tell that my students anticipated the end of my lectures. I wasn't as precise as the clock or the calendar, but I hardly ever went over time. The students always started packing up long before I was through talking. I found it strangely comforting as it was a relief to know I was about done.

Pretty much everything comes to an end. As far as I can tell, it isn't the endings that bother humans. It's their difficulty predicting the endings. That's why the calendar is cool. You know when the year will end, and it's over in an instant.

We all die. Of course, on the other days, we all live. Not a bad trade-off REALLY. So far, I have lived over 25000 days. I am being purposefully vague here, but I only have one day of dying still to come.

I don't like the idea of bucket lists. Having one would make me feel all rushed to get things done. I would be especially ticked off if I got everything done and then found out I had to sit around for years waiting to die. My personal goal is to take things nice and easy up to the day I die and then kind of taper off.

It would be convenient if we knew when we needed to have our affairs in order. Of course, some of us would just put it off to the last minute anyway.

So why, if all living things are made of the same basic stuff, do all have different life spans. How come the tortoise gets 73,000 days and humans get, on average, a third of that? Maybe I need to slow down. . .

It turns out that the difference in lifetimes is because of the number of heartbeats we are allotted. Way back in the 1930's, a Swiss named Max Kleiber studied the relationship between mass and metabolism. He came up with a formula predicting the energy burned per unit of weight is proportional to an animal's mass raised to the three-quarters power. In plain words, the smaller you are, the more calories it takes per ounce to keep you alive.

However, if you eat a lot of food, you have to metabolize it quickly so that you will have room for the more food you are going to eat. Metabolizing food requires energy and generates heat. Getting rid of heat and circulating energy require heartbeats.

It turns out that, at least for mammals, there is an allotted number of heartbeats per lifetime: about a billion. A relaxed shrew, which may be an oxymoron, has a resting heartrate of about 850 beats per minute. At that rate they have a life expectancy of about two years. Some whales have heart rates of ten to fifteen beats per minute. This buys them about two hundred years.

I figure that, on the average, my heart rate has been about seventy beats per minute for most of my life. Of course, there were those occasions when I did stupid, terrifying things that increased the rate. There is also my wife who still takes my breath away. I figure those times are offset by all the hours at work, during which time, I didn't really do much of anything. Seventy is probably a good average.

So, if you calculate this out to a billion heartbeats, I'm already dead. And I have been since I was about twenty-seven years old. Humans seem to beat the billion-heartbeat rule with brains. It's not that we live longer because we live smarter, but it's because we use medicine and interventions to extend our lifespan. I have to admit though, It's not very predictable.

SERIOUSLY - A POSTSCRIPT

What could one possibly say as a postscript to reality? I mean, what comes after reality? A post-reality almost has to be defined as science fiction doesn't it?

ABOUT THE AUTHOR
　　　Dr. Gary McCallister is Professor Emeritus of Biology from Colorado Mesa University where he taught for 42 years. His research area was in Infectious Diseases but he taiught many courses in general science and science education.　He is also an award winning author of a Newspaper column entitled Simply Science.

OTHER BOOKS BY GARY MCCALLISTER

MUSIC
Making More than Music 2014
First Songs With the Mountain Dulcimer: history, instrument, and simple songs 2015
Hymns on Mountain Dulcimer: Learn to play the mountain dulcimer using hymns 2016

SCIENCE
Hanging Out With GRAVITY: Galileo's gravity game 2015
Seriously Silly Science: A science reader for the whole year - and some of it is even true 2015
A Convenient Truce: A cease fire in the war between religion and science 2016
The Solar Solution: the solution to problems you didn't even know you had. 2017
Between Two Mirrors: art and science in the modern world 2017

NOVELS
Walking Man 2015

www.ingramcontent.com/pod-product-compliance
Lightning Source LLC
Chambersburg PA
CBHW070307230526
45470CB00002B/760